Interstellar dust grain: diameter 4×10^{-5} inch

Blue light wavelength: 1.9×10^{-5} inch

Bacterium: diameter 4×10^{-5} inch

Black hole: diameter 40 miles

Large moon crater: diameter 120 miles

Largest asteroid: diameter 620 miles

Mars: diameter 4,217 miles

White dwarf: diameter 5,000 miles

Venus: diameter 7,521 miles

FRONTIERS OF TIME

TIME®
LIFE BOOKS

This volume is one of a series that examines the universe in all its aspects, from its beginnings in the Big Bang to the promise of space exploration.

VOYAGE THROUGH THE UNIVERSE

FRONTIERS OF TIME

BY THE EDITORS OF TIME-LIFE BOOKS
ALEXANDRIA, VIRGINIA

CONTENTS

The grand cycle of the death and rebirth of the universe—a story told in many cultures—is symbolized by a bronze of the Hindu god Shiva *(below, left)* and an Aztec calendar stone. Shiva holds the fire of destruction in his left hand, and the drum of creation in his right. In the Aztec stone, panels depicting four preceding cosmic ages surround a representation of the current world: the age of the fifth sun.

At the end of a thousand cycles of the four Ages, when the surface of the earth is mostly exhausted, a very fierce drought occurs and lasts for a hundred years. . . . The lord Vishnu enters into the seven rays of the sun and drinks up all the waters, leaving none. . . . All the water that he has taken away goes directly, by his authority, to the seven rays of the sun, which, enlarged by those waters, become seven suns. These seven suns blaze above and below, burning the entire triple world . . . and the earth becomes like the back of a tortoise. When Vishnu has become Rudra of the doomsday fire, the one who withdraws everything, he becomes the hot breath of the serpent of Infinity and burns down the subterranean watery hells. And when it has burnt all the hells, the great fire goes to the earth and burns up the whole surface of the earth. Then a most terrible whirlwind haloed in flames envelops the middle realm of space and the world of heaven. . . . When Vishnu . . . has thus burnt the entire universe, he creates clouds that are born out of the breath from his mouth. . . . Roaring loudly, they fill the sky with their huge bodies. Sending down their rain in great torrents, they extinguish the terrible fire. . . . Then a wind comes out of the breath from the mouth of Vishnu, that blows for more than a hundred years and destroys those waters. Then the Lord, the Unthinkable, the existence of all beings, the one of whom all beings are made, who has no beginning, drinks up that wind, leaving nothing.

—Vishnu Purana
Tenth century AD

ust as cultures around the world have invented creation myths to explain the universe's beginnings, many—such as the ancient Hindus—have attempted to foretell its end as well. Although the details obviously vary, the picture is remarkably similar from story to story, perhaps reflecting the common human experience of such natural cataclysms as fire, flood, and storm. In the modern world, the study of the end of days has become a more systematic, if no less reverential, pursuit. Its new standard-bearers are the cosmologists—astrophysicists who have devoted themselves to examining the origin, evolution, and eventual fate of the universe.

Equipped with state-of-the-art investigative techniques and a sophisticated understanding of the workings of the physical world, the twentieth century's most imaginative scientific minds continually probe to the outer limits of reality in their quest for explanations. Yet for all their successes, an air of mystery still surrounds key aspects of the story; science has found that there are events that lie beyond its ability to describe.

As far as Earth's future and ultimate demise are concerned, however, the view is much clearer and, it turns out, hauntingly familiar. Just as the Hindu sages prophesied, the end of the world will begin with fire and drought. Steadily rising solar radiation will evaporate Earth's waters over a period of a few billion years. About five billion years from now, when the Sun has exhausted its store of thermonuclear fuel, it will swell into a vast red giant star that will transform the planet first into a desiccated cinder and then into a ball of molten rock. Tossing off its outer atmosphere in a raging wind of super-energetic particles, the dying Sun will then collapse under the force of its own gravity, shriveling to a tiny, dim white dwarf and leaving behind a collection of frozen planetary corpses veiled in perpetual darkness. Even the Hindu time scale is uncannily accurate. The great dissolution described in the quotation above takes place at the end of a single day in the life of Brahma—one manifestation of the supreme being in the Hindu pantheon—which works out to more than eight and a half billion years, not too far from the Sun's true lifetime of about ten billion years.

Of course, modern science looks to more distant frontiers. Astronomers estimate that the universe has been expanding for some 15 to 20 billion years, ever since its presumed birth in a colossal explosion known as the Big Bang. In fact, they say that the universe could conceivably go on expanding for-ever—until all the stars have burned out, until the galaxies have transformed themselves into vast black holes, until matter itself decays into a thin haze of elementary particles, until everything that now exists has faded away into nothingness. But according to current information, an opposite outcome is equally possible, and here again the ancient wisdom offers insights. Although the Hindu mystics had no counterpart for the expansion of the universe, they did envision a cosmos that goes through endless cycles of birth, death, and rebirth, a view shared by some theoretical physicists. If the universe contains enough matter—a proposition neither confirmed nor denied so far by obser-vations—one day in the remote future its expansion will slow to a stop. Then, in a reversal of the Big Bang known as the Big Crunch, the cosmos will begin to contract until it finally collapses to a point of infinite density. Because physics cannot describe such a state, what happens next is pure conjecture, but a few daring cosmologists, spiritual heirs to the first mythmakers, imag-ine that, just as Brahma opens his eyes each day on a whole new universe, the process could commence all over again with a new Big Bang, and the cycle could repeat forever.

Perhaps most difficult to grasp in these strange tales of beginnings and endings are the huge stretches of time involved. Galaxies, for example, will take not a billion, or a billion billion, but a billion billion billion years to die. If the universe never stops expanding, interactions between certain sub-atomic particles may still be occurring at a time equal to a galaxy's duration repeated a billion billion billion times, then a billion billion billion times more, then a billion billion billion times more than that. The basic chronological unit in such scenarios is itself incomprehensible in human terms. A billion years ago the only creatures existing on Earth were one-celled organisms, and much of evolution's rich tapestry of life was yet to unfold. A billion years hence Earth will have altered completely, wrought anew by continuously shifting continents, recurrent ice ages, and occasional cosmic bombardments by asteroids and comets. Against such an endless backdrop, the entire 5,000-year history of civilization has been no more lasting than the flash of a lightning bug.

EARTH IN FLUX

During its brief moment under the Sun, humanity has received a misleading impression of the planet, its mountains and oceans seeming like permanent aspects of an essentially stable countenance. In reality, the only constants as time progresses are cycles of incessant change. Nowhere is this more apparent than in an examination of Earth's future, informed by analyses of its past. Actually, investigators of the geologic record have been relatively slow to recognize just how mutable Earth is. Geologists have known since the middle 1800s that ice ages play a crucial role in sculpting the landscape, but not until the 1960s did they wholeheartedly accept the notion that the continents themselves are variable features of Earth's surface, drifting across the globe to smash together and then rip apart again in a kind of slow-motion demolition derby that shows no signs of letting up.

Earth's surface consists of about ten so-called tectonic plates, on which the continental landmasses rest. As these plates slide over the partially molten layer below, they separate along midoceanic ridges, allowing magma to well up and form new crust. By analyzing patterns in the spreading of the seafloor along these ridges and by studying comparable clues to ancient magnetism locked in the geologic record, researchers have been able to trace plate movements back nearly 600 million years. About 514 million years ago, for example, not long after the first shelled creatures appeared, according to the fossil record, the expanse of rock that would later be named North America lay along the equator, with the present-day East Coast facing south, and fragments of what is now the Eurasian continent were scattered around the globe. At the same time, the individual continents now called Australia, Antarctica, Africa, and South America, as well as parts of the Middle East and southern Asia, were welded into a vast landmass geologists have named Gondwana.

More than 200 million years later, Gondwana collided with North America, having drifted southward all the way across the South Pole and then northward again to the equator. The collision brought together what are now the East Coast of the United States and the Moroccan coast of Africa, with the resulting pileup raising a mountain range as high as the modern Himalayas. The eroded roots of this range are now called the Appalachians. By the dawn of the age of dinosaurs, about 250 million years ago, most of the major continents had joined into one sprawling supercontinent known to geologists as Pangaea. But as with everything else about the planet's chameleon-like surface, Pangaea was only temporary; within about 100 million years it started to break apart, its pieces eventually scattering to form the continents of the modern world.

The motion continues at a slow but inexorable pace. Seafloor spreading at the Mid-Atlantic Ridge, for example, indicates that North America and Europe are drifting apart at the rate of about one inch per year. Minimal as this seems, the effects are perceptible. During the 1960s and 1970s, new islands emerged from the ocean near Iceland as a result of underwater extrusions of magma that rose to fill the rift along the ridge. When plates separate within a continental landmass, as in the Great Rift Valley of Africa, lava welling up through volcanoes serves to rebuild the crust. And the devastating earthquakes that plague countries all around the Pacific Ocean are an ever-present reminder of the grinding of plates against each other.

The next 50 million years of unremitting movement will see the geographic map transformed as dramatically as the political map has been in the last 50 years. If present movement trends continue, California, sitting on the boundary of the Pacific and North American plates, will have ripped apart along the San Andreas fault, and Los Angeles will have passed west of San Francisco and be approaching Alaska. The Atlantic and Indian oceans will have grown at the expense of the Pacific. The eastern coast of Africa will have split off along the Great Rift Valley to form a new island continent, and the remainder of Africa will have plowed north into Europe, squeezing out the Mediterranean Sea. By 150 million years from now, according to one scenario, the Eurasian-African landmass will have rotated clockwise, and Australia and Antarctica will once again have merged. Beyond this point, so many variables influence the outcome that the pattern becomes impossible to predict.

Earth scientists are still not completely sure what drives this unceasing flow of the tectonic plates, but in general terms they agree that the motions arise in the planet's partially molten interior, as upwelling heat causes hot, buoyant rock to rise and fall in huge eddies, rather like soup simmering on the stovetop. The plates—in effect, the cooled parts of enormous convection cells—simply ride along on top. Because the plates are rigid and the convective patterns are orderly, plate movement is not haphazard. For example, in places known as subduction zones, the edge of one plate will slip under another and be dragged down into the interior, pulling the rest of the plate with it in a steady direction.

MAGNETIC IRREGULARITY
Deeper below the surface, the roiling action within a fluid layer known as the outer core may lead to another form of instability, in Earth's magnetic field.

For more than 3.6 billion years, Earth has undergone a dynamic evolution exemplified by the bubbling up of molten rock from the planet's interior. Eruptions such as this one in Zaire's Nyiragongo volcano occur several dozen times a year, usually at boundaries between the large plates that make up Earth's crust, reshaping the face of the planet by creating new mountains, islands, and plains.

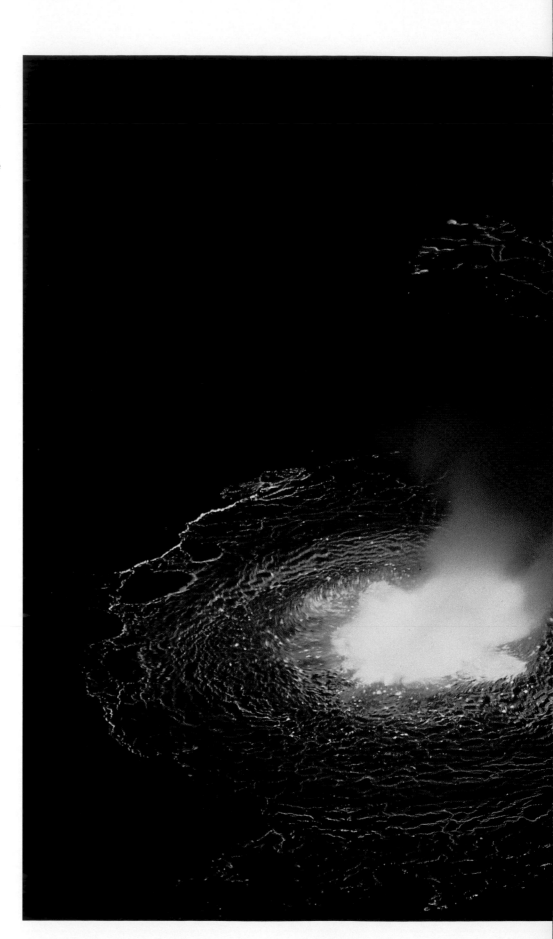

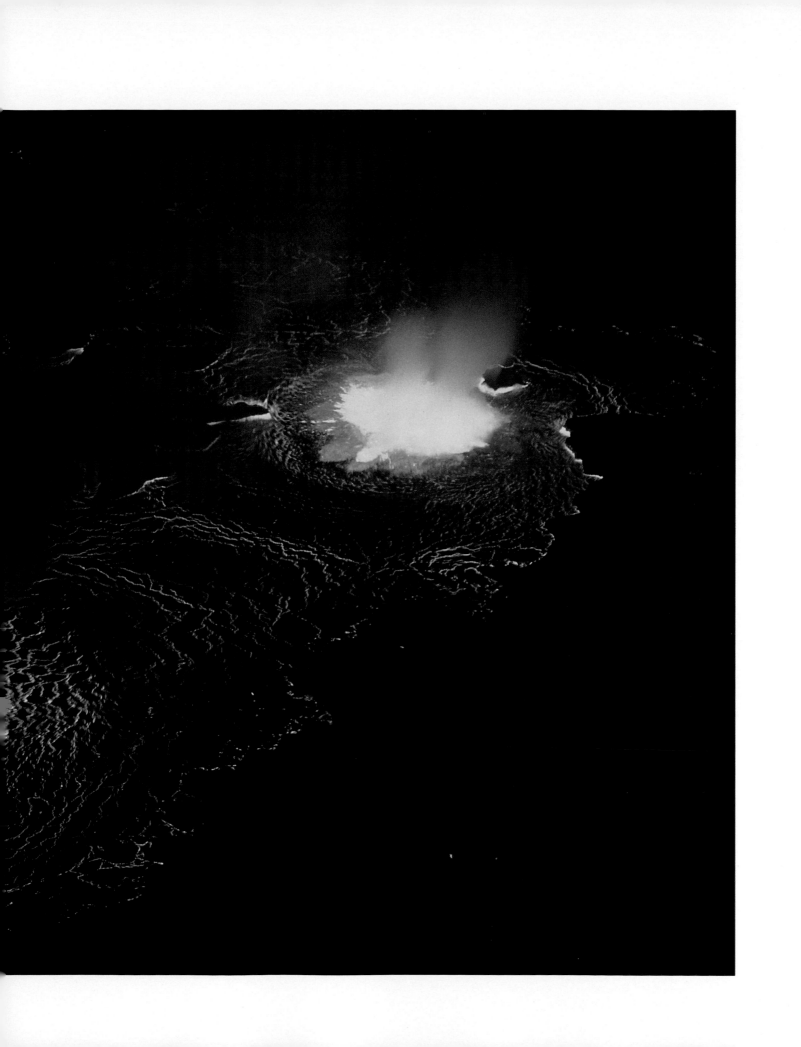

According to accepted theory, the region of magnetic influence that surrounds the planet, typically represented as lines of magnetic force that loop out into space, results from the generation of electric current in the highly conductive molten metals of the outer core as they flow around an inner core of solid iron—producing, in effect, a planetary dynamo *(pages 18-19)*. The magnetic field's overall alignment assumes the classic dipolar shape produced by an ordinary bar magnet, with field lines emanating from near the South Pole and returning to the North Pole. Since the surging of material near the core is caused in part by Earth's rotation, most researchers believe that the field has always been and will continue to be generated, but evidence shows that neither its strength nor its orientation, or polarity, has been consistent. For instance, 730,000 years ago its polarity was reversed: North-seeking compass needles would have pointed south. In fact, examination of the orientation of magnetic minerals frozen into ancient lava reveals that the field direction has switched back and forth many times in the past.

Scientists can only speculate as to what causes these reversals. According to one scenario, the upwelling of material in the outer core could produce small patches of magnetic field lines, or flux, whose polarity is opposite that of the surrounding region. Over time, these patches of reverse flux could weaken and disturb the dipolar component of Earth's magnetic field until it disappeared altogether. Paleomagnetic evidence seems compatible with this view. Isolated regions of opposite polarity have been identified, and other signs indicate that every so often, the intensity of the dipolar field has dwindled to as low as 10 percent of its full strength, remaining in a weak and unstable transitional state for anywhere from 5,000 to 10,000 years. When the

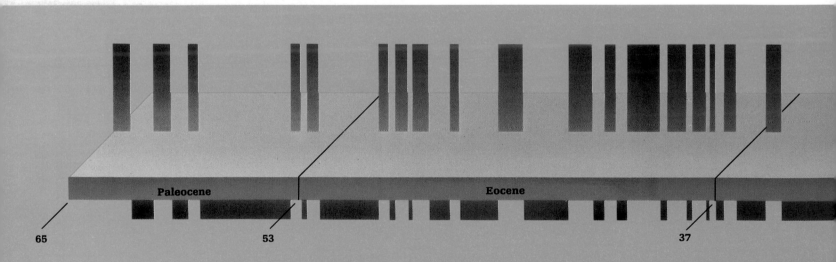

Paleocene   Eocene

65   53   37

EARTH'S MAGNETIC HISTORY

The discovery in 1906 that some rocks are magnetized with an orientation, or polarity, opposite that of Earth's present-day magnetic field eventually led scientists to conclude that the field reverses from time to time. They have since

field regains full dipolar strength, it does so sometimes with polarity unchanged, other times with its poles reversed.

Intervals between actual reversals are quite irregular, ranging from 35 million years to several thousand years. Predicting the next reversal is thus little better than a game of chance, but there are indications that dipolar decay has already begun. During the past 150 years, the magnetic field has declined in strength by 10 percent. If the decline continues at this rate, the field will reach zero in about 1,500 years.

Such an event is of more than scholarly interest. Perhaps the most drastic impact would come from solar cosmic rays, ultra high energy protons spewed from the Sun during periodic, gigantic eruptions known as solar flares. The most violent flares can increase the radiation from the Sun's surface tenfold and deliver lethal doses of cosmic rays to unshielded astronauts. Today, Earth's magnetic field deflects the vast majority of these destructive particles, although a few spiral down the field lines at the poles, entering the atmosphere and triggering chemical reactions between atmospheric nitrogen and oxygen. The resulting nitrogen-oxide compounds tend to damage the protective ozone layer in the upper atmosphere that shields living things from the harmful effects of the Sun's ultraviolet radiation. During a magnetic decline, the onslaught of solar flare particles would be more global *(pages 20-21),* with potentially fatal consequences for terrestrial organisms.

RELENTLESS ICE
While the never-ending drift of the continents and fluctuations of the magnetic field may be traced to the vagaries of Earth's interior, other patterns of

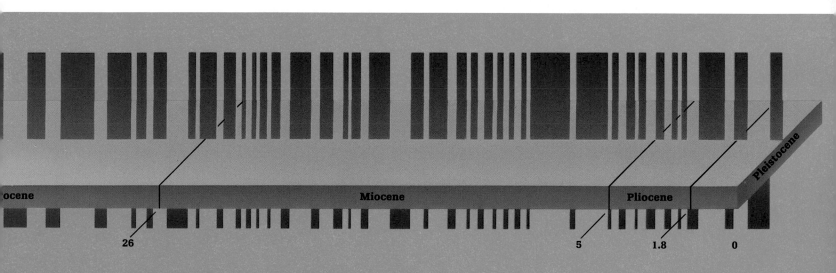

ocene

Miocene

Pliocene

Pleistocene

26

5

1.8

0

determined that the shifts occur at irregular intervals, ranging from a few thousand to 35 million years. The last reversal is thought to have been about 730,000 years ago, and there are signs that Earth may be undergoing a transition to the next changeover *(pages 17-21).* In the time line above, periods of polarity corresponding to Earth's present field direction are indicated by bars rising above the central plane; descending bars represent times of reverse polarity.

change may owe to influences of more cosmic proportions. Most notably, the global climate has alternated between periods of relative warmth and episodes of extreme cold, when snow and ice cover up to 30 percent of the planet's surface. All sorts of geologic data—from fossil records and signs of glacial scouring to variations in the thickness of ancient antarctic ice—reveal consistent cycles of waxing and waning ice epochs over the last 500,000 years, some more intense than others. Typically, the ice advances from the poles toward the middle latitudes for tens of thousands of years and then, for 10,000 years or so, it retreats while the world bathes in a so-called interglacial period. Roughly 140,000 years ago, for example, palm trees and other tropical plants grew in northern Europe, while Greenland, Alaska, and northern Norway luxuriated in temperate-zone forests of oak, maple, and birch. But during the last ice age, which peaked about 25,000 years ago and ended little more

As often as several times a day, the Sun emits gases in violent outbursts known as eruptive prominences, seen here in an extreme ultraviolet spectroheliogram taken by Skylab in 1974. Such eruptions may be accompanied by a shock front that accelerates particles in interplanetary space to such speeds that they can damage Earth's protective layer of atmospheric ozone—a possibility that increases when the planet's magnetic field is undergoing a reversal of polarity.

A REVERSIBLE CLOAK

Among the changes that may await Earth in the relatively near future is a reversal of the orientation of its magnetic field, so that the north end of a compass needle will point to geographic south rather than geographic north. The geologic record indicates that many such reversals have occurred in the past and another flip-flop may already be under way, but geophysicists are only just beginning to understand the complex processes involved.

Earth's magnetic field originates deep within the planet *(pages 18-19)* and extends thousands of miles into space, forming a kind of cosmic cloak, known as the magnetosphere *(below)*. Much as a windsock billows or droops in response to the strength of the breeze, the magnetosphere is shaped by the pressure of the solar wind, a continuous flow of charged particles and embedded magnetic field generated by the Sun. As shown here, the magnetosphere generally maintains an overall bullet shape, blunt on the planet's sunward side and streaming out behind in a feature called the magnetotail.

Ordinarily, when Earth's magnetic field is in a stable configuration known as the dipolar state—much as it is today—the solar wind is largely deflected by the magnetosphere at the so-called bow shock *(green)*. Solar particles *(pink)* breach the defense only at the north and south magnetic poles, where solar magnetic field lines break and reconnect with terrestrial field lines *(light blue)*. When the field is transitional, other gaps appear *(pages 20-21)*, and Earth becomes more susceptible to dangerous solar messengers.

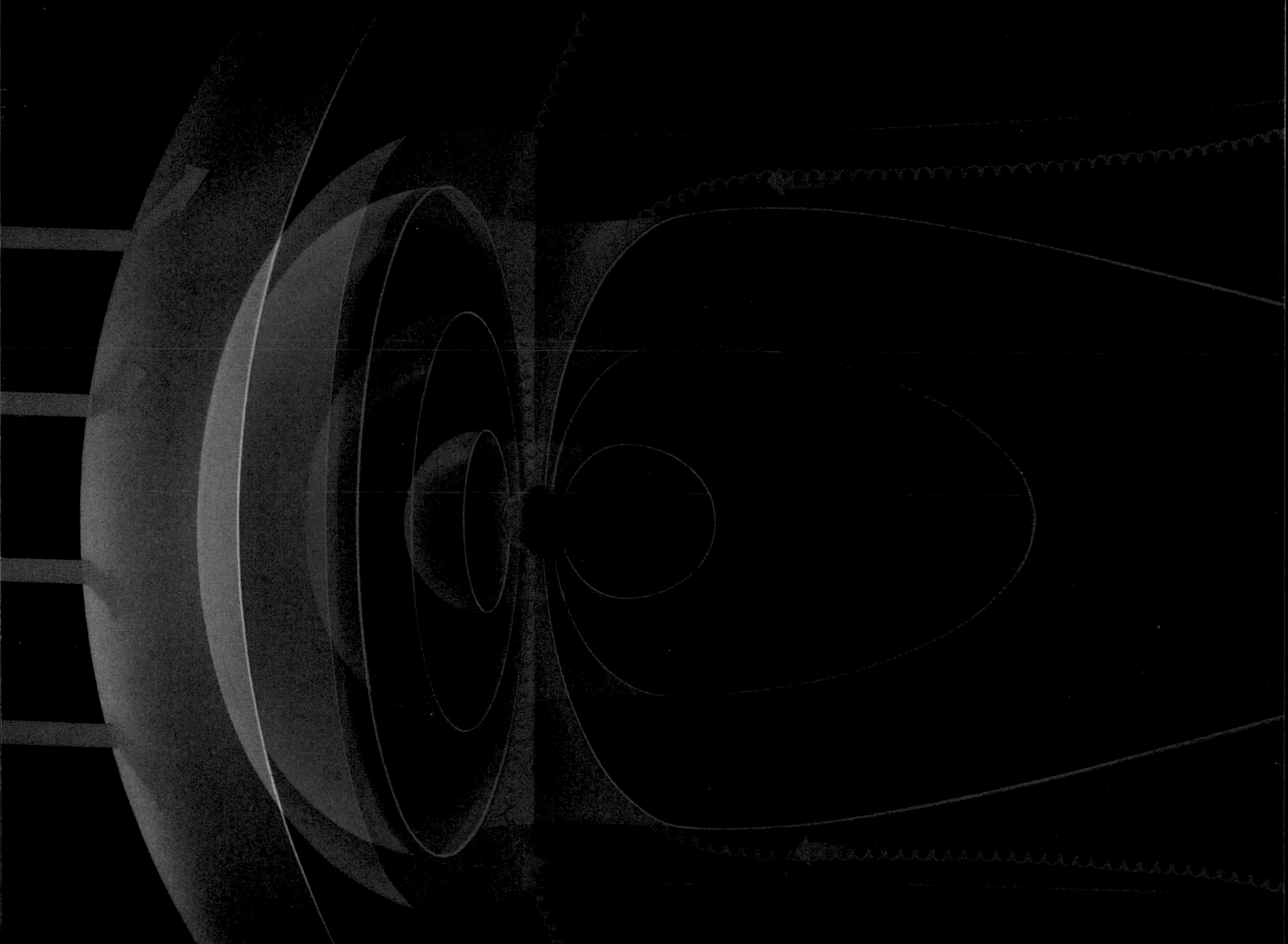

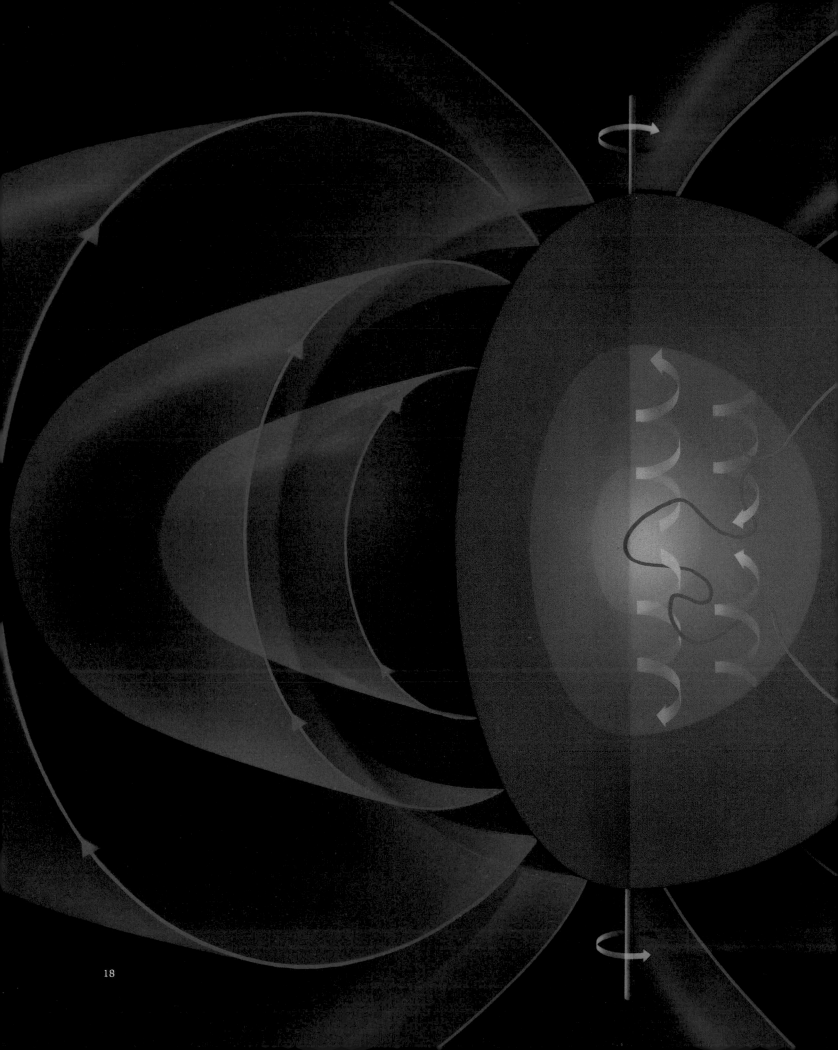

18

WORKINGS OF A PLANETARY GENERATOR

The most widely accepted theory of the origin of Earth's magnetic field suggests that the planet is, in effect, a self-exciting dynamo *(left)*, generating electricity through complex interactions among a thick layer of molten iron alloy known as the outer core *(orange)*, a solid iron inner core *(yellow)*, and the planet's rotation. That electric current, in turn, generates a globe-girdling magnetic field.

The action of this geodynamo creates field lines of force, often called magnetic flux, in a pattern remi-niscent of the field lines from a simple bar magnet. On present-day Earth, the flux lines emerge into space from the Southern Hemisphere and reenter the planet in the Northern Hemisphere. At the boundary between the outer core and a less dense upper layer known as the mantle *(brown)*, the magnetic field mirrors the shifting patterns of convection in the molten outer core. As a result, patches of reverse flux—regions where some field lines are oriented opposite the lines surrounding them—occur sporadically in both hemispheres *(below)*. Some scientists theorize that the gradual propagation and intensification of reverse-flux patches may periodically trigger a planetwide reversal of the magnetic field.

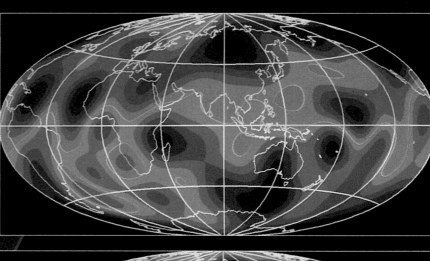

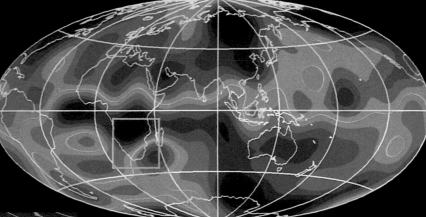

The two maps above, projected from measurements made at the surface in 1777 *(top)* and 1980 *(bottom)*, show the directional orientation of Earth's magnetic field at the core-mantle boundary. Patches of reverse flux—inward *(blue)* in predominantly outward *(red)* regions and vice versa—may be the result of upwellings in the fluid outer core, as shown at left. Magnetic field lines associated with upwelling fluid *(orange)* diffuse up and down through the mantle *(brown)*, producing localized spots of outward or inward flux in opposition to the flux direction of the surrounding region.

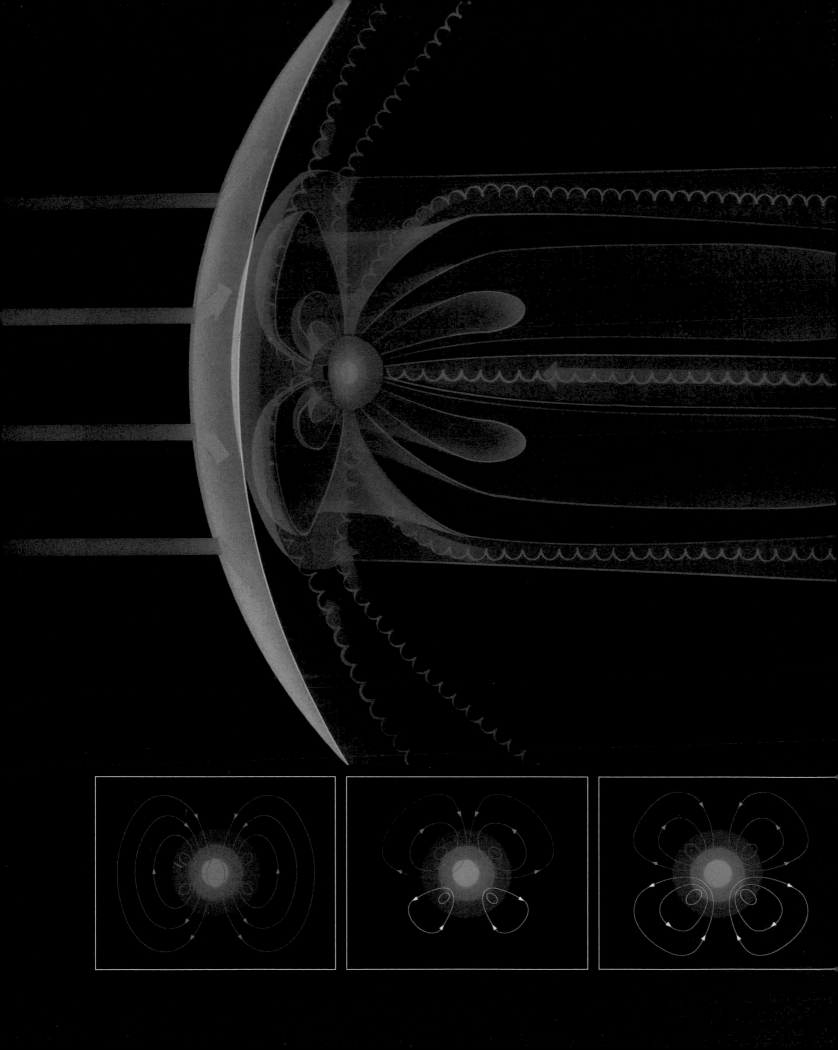

A Threadbare Magnetosphere

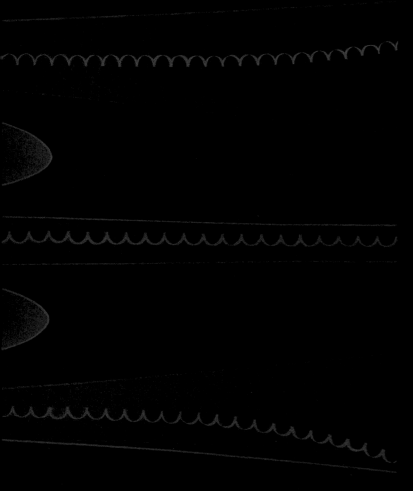

Preliminary signs of a magnetic field reversal include a marked decrease in field intensity, measured in units known as nanoteslas. The strength of today's field, for instance, seems to be decreasing by about seven percent each century. As the decline progresses, the stable dipolar arrangement begins to give way to more complex—and weaker—field structures. The illustration at left, for example, depicts one hypothesis for the advanced stages of such a transition, the so-called quadrupole configuration.

Not only is the quadrupole magnetosphere less extensive than its dipolar counterpart, but shifting regions of reverse flux *(diagrams below)* allow solar field lines to reconnect with terrestrial field lines at the equator as well as at the poles. This increases access by solar particles to Earth's upper atmosphere, with potentially serious effects. During periods of heightened solar activity, such as the onset of violent solar flares, the Sun releases streams of highly energetic particles called solar cosmic rays that have been linked to the formation of nitrogen oxides. These in turn can cause partial destruction of the atmosphere's ozone layer, which normally prevents harmful ultraviolet rays from reaching the surface. With Earth's magnetic defenses down, a few intense solar flares could substantially deplete the ozone layer, leading to an increase in environmental stress—and possibly the extinction of many of the planet's life forms.

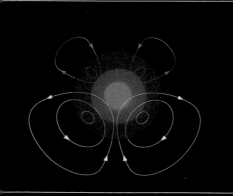

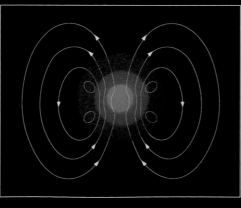

In stage one of the so-called flooding model of magnetic field reversal, Earth's present-day field shows lines looping from the Southern into the Northern Hemisphere *(far left)*. In stage two, a region of reverse flux *(yellow)* appears in the Southern Hemisphere and begins to spread toward the equator. In stage three, field lines exit at the equator and loop into both the north and south magnetic poles, creating a quadrupole magnetic field. In the fourth stage, the reverse flux region expands north of the equator until it has flooded the entire field in stage five. This polarity will dominate until new regions of reverse flux emerge to return the field to the opposite orientation.

than 10,000 years ago, glaciers extended from central Canada as far south as the state of Missouri; all told, ice covered nearly three-quarters of North America and half of Europe.

Researchers in the late 1800s and early 1900s were divided as to the cause of these glacial eras, although the basic mechanism seemed obvious: Ice ages were brought on by the failure of the snows to melt in summer, leading to a steady accumulation and spread of icy material year after year; and they happened because either winters had been too cold or summers not warm enough. Then in 1920, Yugoslavian mathematician Milutin Milankovitch published a theory that not only confirmed the cool-summer scenario but also linked ice age cycles to Earth's orbital behavior. Milankovitch demonstrated that three different features of Earth's orbit yield periodic summertime reductions in the amount of solar radiation reaching Earth, in patterns that are a rough but convincing match for actual ice age frequency. First, the tilt of Earth's axis relative to its orbital plane changes by three degrees in the course of 41,000 years; when the tilt is less severe, summers tend to be cooler in both the Northern and Southern hemispheres because both are less directly pointed toward the Sun. Earth's axis also wobbles, on a cycle of 26,000 years, which causes summer to occur at different points in Earth's orbit—from perihelion, when Earth is closest to the Sun, to aphelion, when it is farthest away. (This pattern, of course, means summers are cool in one hemisphere but warm in the other, explaining why ice ages often afflict only one-half of the globe.) And finally, Earth's orbit itself varies over 100,000 years from elliptical to nearly circular and back, changing the distance to the Sun by more than 11 million miles and thus leading to overall cooler or warmer epochs.

ICE IN THE FORECAST

These cycles interweave in such a complex fashion—sometimes reinforcing and sometimes offsetting one another—and so many other factors have been shown to play a role that the match-up with actual ice ages is not always precise, and predictions can be tricky. Nevertheless, most signs indicate that the Northern Hemisphere's next ice age could be well under way between 3,000 and 7,000 years from now. For example, meteorologist John E. Kurtzbach of the University of Wisconsin has calculated that about 9,000 years ago, around 7000 BC, Earth was receiving about seven percent more solar radiation during midsummer than it does currently. This was not too long after the end of the last ice age, during a period when two of the orbital factors noted by Milankovitch coincided: Earth was making its closest approach to the Sun during the northern summer, and its axis was at maximum tilt. The resulting warming effects prevailed for about 5,000 years, with temperatures in many areas of the globe averaging close to five degrees Fahrenheit warmer than now. But as the alignment of orbital factors shifted, the more northerly latitudes began to grow cooler and, discounting short-term variations, conditions have been on a steady decline for the last 4,000 years.

The shift to ice age conditions could be relatively sudden, however, at least in small areas. Genevieve Woillard, a Belgian botanist, has examined the pollen from layers of peat some 115,000 years old in a bog in Alsace, in northeastern France. In the oldest strata, she found the pollen of trees that flourish in a temperate climate: firs, oaks, and alders. During the next two centuries, spruces and pines, which are characteristic of a cooler climate, gradually started encroaching. And then quite abruptly, in a period perhaps as short as 20 years, the temperate trees disappeared entirely, marking the end of an interglacial period.

Of course, humans have made the world a very different place during their brief reign, and whatever influences produced such a rapid transition 115,000 years ago may prove less telling this time around. On the one hand, the burning of fossil fuels, by increasing carbon dioxide levels in the atmosphere, may be leading to a so-called greenhouse effect, trapping heat and raising global temperatures more than other factors are lowering them. On the other hand, much of what civilization has done to the natural environment may be encouraging heat loss. Throughout North America, Europe, and the Soviet Union, for example, vast areas of forest have been cleared for both timber and agriculture; in satellite photographs taken during winter, these exposed tracts show up as patches of brilliant white that reflect up to 70 percent of the sunlight falling on them—compared with 10 to 20 percent for dense forests—and keep heat from building up. How these competing forces will affect the timing of the next ice age is impossible to predict.

When the ice does come again, though, the results will be dramatic. Based on data from the evidence of previous encroachments, the ice will advance at a rate of between 200 and 400 feet per year. Over the course of centuries, then, most of the great cities of the industrialized world will be ground to rubble, subjected to pressures measuring in the tens of tons per square foot; they will eventually lie buried under glacial ice a mile thick. Whole nations will be obliterated. The sea level will fall as more and more water is tied up in glaciers, thus baring great expanses of coastal plain. Whatever remains of the temperate zones, currently home to most of the world's population, industry, and agriculture, will be heavily glaciated, with a climate resembling that of present-day Lapland or arctic Canada. Meanwhile, the equatorial regions will be burdened with prolonged drought. As the oceans grow cooler and the ice spreads from the north, less moisture will evaporate from Earth's surface; less water vapor in the atmosphere will lead to decreasing rainfall. Ultimately, this drying of the world will help bring the glaciation to a halt by reducing winter snowfalls, but for tens of thousands of years, the summer monsoons of Africa and Asia—crucial to the survival of such staple crops as rice—may very well fail repeatedly.

The full impact on humanity can hardly be imagined, although the severity of the effects will probably depend on how quickly the changes commence. With thousands of years to prepare for a slowly evolving crisis, the world's peoples could develop technological and social responses that might mitigate the effects and usher in an era of unprecedented global cooperation.

But if climatic catastrophes strike so suddenly that unsuspecting populations have to scramble desperately to survive, the next ice age could leave billions dead and civilization in ruins. All of human accomplishment could turn out to be an interglacial phenomenon, and the species might have to start virtually from scratch again.

A CELESTIAL BARRAGE

By all accounts, another type of planetwide cataclysm lies in Earth's future, one that is likelier to catch the world unawares and will spell a swifter doom for many forms of life. Even more directly than in the case of ice ages, and perhaps in more ways than one, the heavens are again to blame.

Scientists learned quite recently just how unpredictable such an event may be. Early in April 1989, astronomer Henry Holt of Northern Arizona University was examining a pair of photographs that he had taken a few nights before through an eighteen-inch telescope on Palomar Mountain. Holt was suddenly brought up short when he noticed that one of the "stars" in the images had moved across the frame in the fifty-minute interval between one photograph and the next. After estimating the object's presumed orbit and working backward from the night the images were taken—March 31—he and other astronomers quickly realized that the star was actually an unknown asteroid that had passed within half a million miles of Earth just eight days before, on March 23.

In cosmic terms it had been an extremely close call. Asteroid 1989 FC, as it was named, was a ball of rock perhaps half a mile wide and weighing several million tons that hurtled along at 44,000 miles per hour. It had crossed Earth's orbit at a point that the planet itself had occupied just six hours before. Had the two collided, the impact would have been equivalent to the explosion of 20,000 one-megaton hydrogen bombs. The resulting crater would have been five to ten miles across and about a mile deep. Depending on where the asteroid hit Earth, the blast wave could have killed many tens of millions of people and perhaps hundreds of millions. And it would have struck without any warning whatsoever.

Although such an unlucky crossing of celestial paths may seem rare, the geologic record suggests that in fact, Earth has frequently been the target of cosmic bombardments by objects of significant size. Researchers studying aerial photographs and satellite images have already identified at least 120 ancient impact craters around the world, and they are finding new ones at the rate of five or six each year. Furthermore, many scientists now believe that the impact of a large asteroid or comet was responsible for one of the greatest catastrophes in the history of life on Earth: the sudden disappearance of the dinosaurs and more than half of the other plant and animal species about 66 million years ago.

In 1980, geologists found strong evidence of such a collision in a thin layer of clay that seems to have been laid down all over the world at the time of the mass extinction. The clay contains high levels of iridium, a platinum-like

The changes wrought by glaciers usually take place over scores or even hundreds of years, as the ice mountains slowly inch forward or back. In 1986, however, the 200-foot-high Hubbard Glacier in Alaska *(below)* made a phenomenal advance, traveling 2,000 feet in two months and cutting off Russell Fjord from Disenchantment Bay. The fjord rose 82 feet and eventually broke through the glacier dam, flooding the region below.

metal that is typically rare in Earth's crust but relatively abundant in asteroids and meteorites. In addition, the clay shows traces of soot, which tell a story of raging, worldwide forest fires. And it contains innumerable tiny grains of shocked quartz, a mineral that can be formed only in the high pressures generated by explosions. Although some scientists think that the body responsible for the iridium layer would have left a crater at least a hundred miles wide, other researchers have set lower limits. One candidate crater—a ring twenty-five miles wide and 66 million years old—lies not far from the town of Manson, Iowa, under several hundred feet of rubble left behind by the last glacier. The asteroid that made it might have measured about five miles across.

THE SCARS OF BOMBARDMENT

Like the rest of the Solar System, Earth has suffered the violent blows of comets and asteroids countless times in its life. Meteoroids of varying size enter the atmosphere millions of times a day, and hundreds of them each year survive to make impact. Often the signs of these bombardments are hidden under the ocean or, if the strikes occurred on land, are soon eroded by wind and rain. Some are still visible, however. Shown here are 4 of about 120 impact sites around the world that bear witness to Earth's vulnerability to external forces.

Gosses Bluff *(right)* in Australia is a circular configuration 13.5 miles across and about 142 million years old. Such formations of uplifted rocks typically occur at the center of complex impact craters.

Less than one mile wide, Wolf Creek crater *(above)* lies on the northern edge of Australia's great interior desert. The crater is about two million years old.

Manicougan crater *(left)* in Quebec, Canada, is a heavily eroded, sixty-two-mile-wide crater thought to be about 212 million years old.

Four craters near Henbury Station, Australia, are part of a cluster of twelve in an area one-half mile square. The cavities were created some 4,000 years ago by fragments of a single meteoroid that broke up just before it hit the ground.

Such large, potentially dangerous pieces of interplanetary buckshot normally reside in the asteroid belt between Mars and Jupiter, a wide band of space teeming with chunks of rock and iron that never managed to assemble themselves into a planet. Their orbits are usually stable, but occasionally a collision in the belt or complex gravitational effects exerted by Jupiter's passage will kick sizable boulders in toward the Sun, where, like 1989 FC, they take up orbits that threaten collisions with Earth.

The conglomerations of ice and rock known as comets can also impinge on Earth. Most scientists believe their breeding ground is a vast region known as the Oort cloud that is thought to lie well beyond the orbit of Pluto, surrounding the Solar System like a hollow sphere. The cloud, which was first theorized in the 1950s by Dutch astronomer Jan Oort, presumably contains billions or trillions of frozen bodies, and every so often a few come plunging toward the inner Solar System. Some race through once and are never seen again. But others, like Halley's comet, get drawn in by the gravitational fields of the planets and the Sun, settling into elliptical orbits that bring them back again and again—often to the vicinity of Earth.

Taking all this into account, impact specialist Eugene Shoemaker of the U.S. Geological Survey in Flagstaff, Arizona, has estimated that there are approximately 1,500 known asteroids and comets the size of a mountain or larger that could one day hit Earth. Every one of them is moving at tens of thousands of miles per hour at the very least—a combination of mass and impetus that would release gargantuan amounts of kinetic energy upon impact.

In an effort to understand just what the effects of such a collision would be, meteorologists Ronald Prinn and Bruce Fegley of the Massachusetts Institute of Technology have done extensive computer modeling of a worst-case scenario: a comet about thirty miles in diameter coming in from the Oort cloud and hitting Earth at 100,000 miles per hour. In 1987, they concluded that the behemoth's tremendous momentum would carry it several miles into Earth's crust, an impact that would release as heat the huge store of energy the comet accumulated during thousands of years of falling toward the Sun. The blast would carve out a crater hundreds of miles across, throwing pulverized rock and other debris into the atmosphere at five times the speed of sound, and heating the air over an area as large as Africa to 3,100 degrees Fahrenheit.

Atmospheric nitrogen and oxygen would literally burn, producing a dense red smog of nitrogen oxides. Within a few hours, the debris plume would arc out into space and then begin to fall back, enveloping the planet in enough dust to create an inch-thick layer over the atmosphere. For three months, until the dust settled, the world would be shrouded in darkness. Meanwhile, beneath this blanket, the nitrogen-oxide smog would have begun to spread around the globe; even after a year, the smog would block any weak sunlight that managed to penetrate the thinning pall above. Slowly combining with water vapor, the smog would convert to nitric acid, and the skies would drip rain as corrosive as the acid in a car battery.

Prinn and Fegley designed their study in an attempt to explain what had happened to the dinosaurs. The simulation more than met the challenge. In a world left barren by the lack of sunlight and eaten away by a deadly precipitation, the threat to all forms of life seemed insurmountable. As Prinn put it in discussing their results, "My problem isn't so much in killing things off, but in finding ways for things to survive."

CYCLES OF EXTINCTION

As the belated discovery of the near miss by asteroid 1989 FC demonstrated, a devastating impact is completely unpredictable, at least in the short run. Indeed, most scientists would say that the impacts occur at random intervals. But a vocal minority believes there is evidence that impacts are a regularly repeating feature of the past, even more dependable in their timing than the ice ages, and questions of timing and predictability are actually hotly debated aspects of the issue of a presumed connection between comet impacts and species extinctions.

In 1983, for example, paleontologists David Raup and Jack Sepkoski of the University of Chicago published a paper in which they surveyed Earth's mass extinctions and concluded that these events have occurred periodically, at intervals of about 26 million years. The following year, NASA scientists Michael Rampino and Richard Stothers came at the issue from the opposite side. After studying the ages of impact craters around the world, they found a 33-million-year periodicity, which coincided with a number of geologic upheavals that could feasibly have been triggered by the tremendous jolt of such collisions. The pair then looked at studies of mass extinctions and concluded that those apparent cycles supported a theory of regular episodes of celestial bombardment.

Although geologists are leery of placing undue weight on these seeming relationships—primarily because the dating of ancient events is an inexact science and any demonstration of periodicity relies more on statistical correlations than on clear physical evidence—astronomers have been driven to speculate as to what could cause cyclical disasters on such a long time scale. A few have suggested that the most likely explanation involves the motion of the Sun around the center of the galaxy.

Detailed charting of the distribution of stars and interstellar matter in the Milky Way and analyses of stellar motions have enabled astronomers to determine not only the galaxy's structure but also the Solar System's galactic location and its orbital behavior. Like many other star systems, the Milky Way is a great pinwheel, the majority of whose stars form a series of spiral arms sweeping out from the hub in graceful arcs. These arms, along with most of the galaxy's interstellar clouds of dust and gas, reside in a wide, thin disk known as the galactic plane. The Sun and its planets sit about 27,000 light-years from the center, near one of the spiral arms and slightly above the galactic plane. Orbiting at a speed of about half a million miles per hour, the Solar System takes 250 million years to make a complete circuit. It does not, however, trace a simple flat circle or ellipse as it revolves; its motion is more

Dangerous Passage

The dating of impact craters on Earth suggests that bombardments occur about every 30 million years—a cycle that roughly corresponds with the Sun's periodic crossing of the midplane of the Milky Way as it orbits the galactic center *(right)*. One theory suggests that passage through the dense concentration of gas, dust, and stars in the midplane *(blue)* perturbs the orbits of comets believed to hover around the Sun in a vast and distant swarm known as the Oort cloud *(far right)*. Kicked out of their normal trajectories, the comets fall in a deadly rain on the inner Solar System.

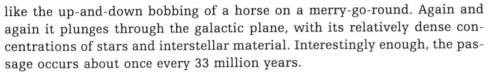

like the up-and-down bobbing of a horse on a merry-go-round. Again and again it plunges through the galactic plane, with its relatively dense concentrations of stars and interstellar material. Interestingly enough, the passage occurs about once every 33 million years.

The correspondence with the apparent extinction and impact cycles seems uncanny. In fact, a number of scientists first noted a connection to extinction patterns back in the 1970s, several years before the discovery of the iridium layer set researchers thinking about the role of impacts. Several of these researchers outlined the factors that could make the Solar System's trip through the galactic plane a perilous one. If the Sun and planets were to pass close to a particularly dense clump of interstellar gas, for example, the clump's gravitational influence could knock loose a shower of comets from the Oort cloud and send them toward the inner Solar System. Starting out slowly, the comets might take thousands or millions of years to arrive, but eventually they would fill Earth's skies with potentially lethal projectiles, any one of which could cause a mass extinction if it hit. Over time, some comets that had taken up solar orbits would break up as their ice evaporated in the heat of the Sun, leaving behind swarms of rocky fragments—many still large enough to inflict grave damage. A number of astronomers suspect that most of the so-called Earth-crossing asteroids, a class of objects whose orbits intersect that of Earth, originated in this way.

As devastating as impacts would be, however, they may not be the foremost hazard. Disintegrating comets would release huge amounts of dust as well as rocks, and although the individual dust grains would be less than a thousandth of an inch across, there would be so many of them filling the inner Solar System that they might seriously dim the sunlight reaching Earth. This interplanetary shroud would last much longer than the debris cloud from an

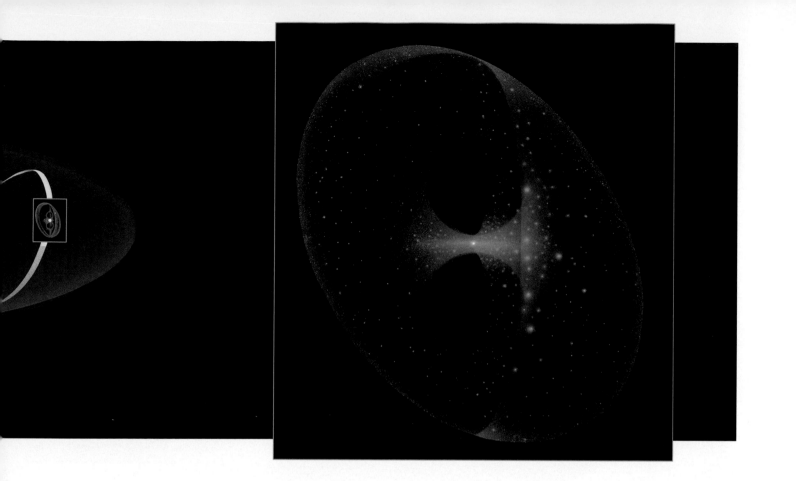

impact: The dust would persist until it was blown away by the solar wind, which could take a few thousand years. Certain long-period ice age cycles might be explained by such a scenario.

Relying as it does on circumstantial evidence, the galactic plane theory remains controversial. If it is correct, however, Earth may be experiencing the consequences of an Oort cloud disruption even now. Judging from its current position in the galaxy, the Sun appears to have emerged about five to ten million years ago from Gould's Belt, a narrow ring of massive B-type stars and dust inclined twenty degrees to the galactic plane and associated with one of the Milky Way's spiral arms. Furthermore, the Solar System seems to have encountered an unusually high number of Earth-crossing asteroids and comets in the recent past. A look at the rate of crater formation both on Earth and on the Moon suggests that the last few million years have been a particularly active phase. In addition, there is some evidence that a giant comet entered the inner Solar System about 20,000 years ago and began to break up, increasing the likelihood of devastating collisions. One fragment may have formed the comet Encke, which has appeared in Earth's skies roughly every three years for the last two centuries, and another may have disintegrated into a dense meteor swarm that struck the Moon in 1975, where it was recorded by seismic instruments left by the Apollo astronauts. Yet even if the correlation between Solar System encounters with the galactic plane and bombardment episodes holds up, individual deviations of millions of years from the presumed cycle make it impossible to gauge the chances of another mass extinction in the near future.

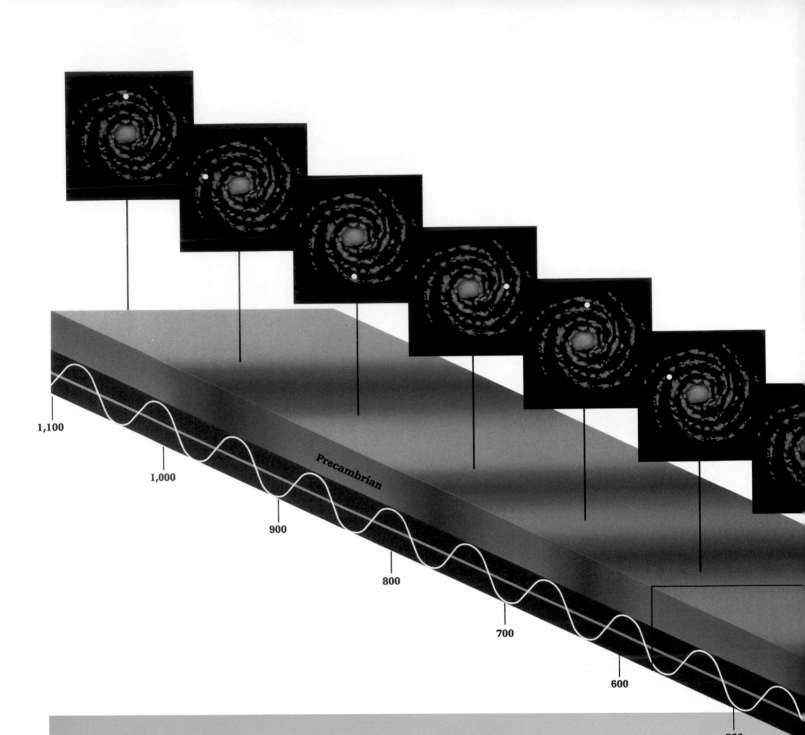

1,100

1,000

900

800

Precambrian

700

600

500

CONNECTIONS OR COINCIDENCE?

In recent years, scientists have looked at possible links between the Sun's movement through the galaxy and geophysical events and processes on Earth. For instance, as depicted in the series of galaxy images above, about once every 100 million years, the Solar System crosses one of the Milky Way's spiral arms, regions about 1,500 light-years wide that are dense with gas and dust. Some investigators, theorizing that passage through a dusty arm could change the amount of sunlight reaching Earth, point to a possible correlation with the occurrence of long ice epochs *(dark purple).* Others liken the effects of this passage to those brought on by the Sun's crossing of the galactic plane *(yellow curve)*—that is, a periodic hail of comets. Still others note that because spiral arms are the breeding ground of massive, short-lived stars that expire as violent supernovae, Earth would have more frequent encounters with the high-speed particles—cosmic rays—ejected by the explosions, which could lead to partial destruction of the planet's protective ozone layer.

DEATHS IN THE STELLAR NEIGHBORHOOD

As epic as the cycles of impacts and extinctions appear to be, other disturbances, geared to even lengthier time scales, may have plagued Earth's past and certainly haunt its future. The agents of destruction are some of the more familiar beacons in the night sky.

One easily recognizable inhabitant of the winter heavens is Betelgeuse, the bright red star that adorns the right shoulder of Orion the Hunter. A well-known lantern since ancient times, Betelgeuse is actually a celestial bomb waiting to go off. Many times more massive than the Sun, it is a huge, bloated star of a type known to astronomers as red supergiants. Sometime in the comparatively near future—tomorrow, perhaps, or a few thousand years from now—the star will finally exhaust all of its thermonuclear fuel. Following that, in the space of less than one second, Betelgeuse's core will collapse into a tiny, incredibly dense neutron star only about twelve miles across, while its upper layers are blown outward at nearly one-tenth the speed of light in a cataclysmic explosion known as a supernova. For a few weeks it will shine with the light of many billions of Suns. From front-row seats only 650 light-years away, humankind will be bathed by its glow, as bright as the light of the full Moon.

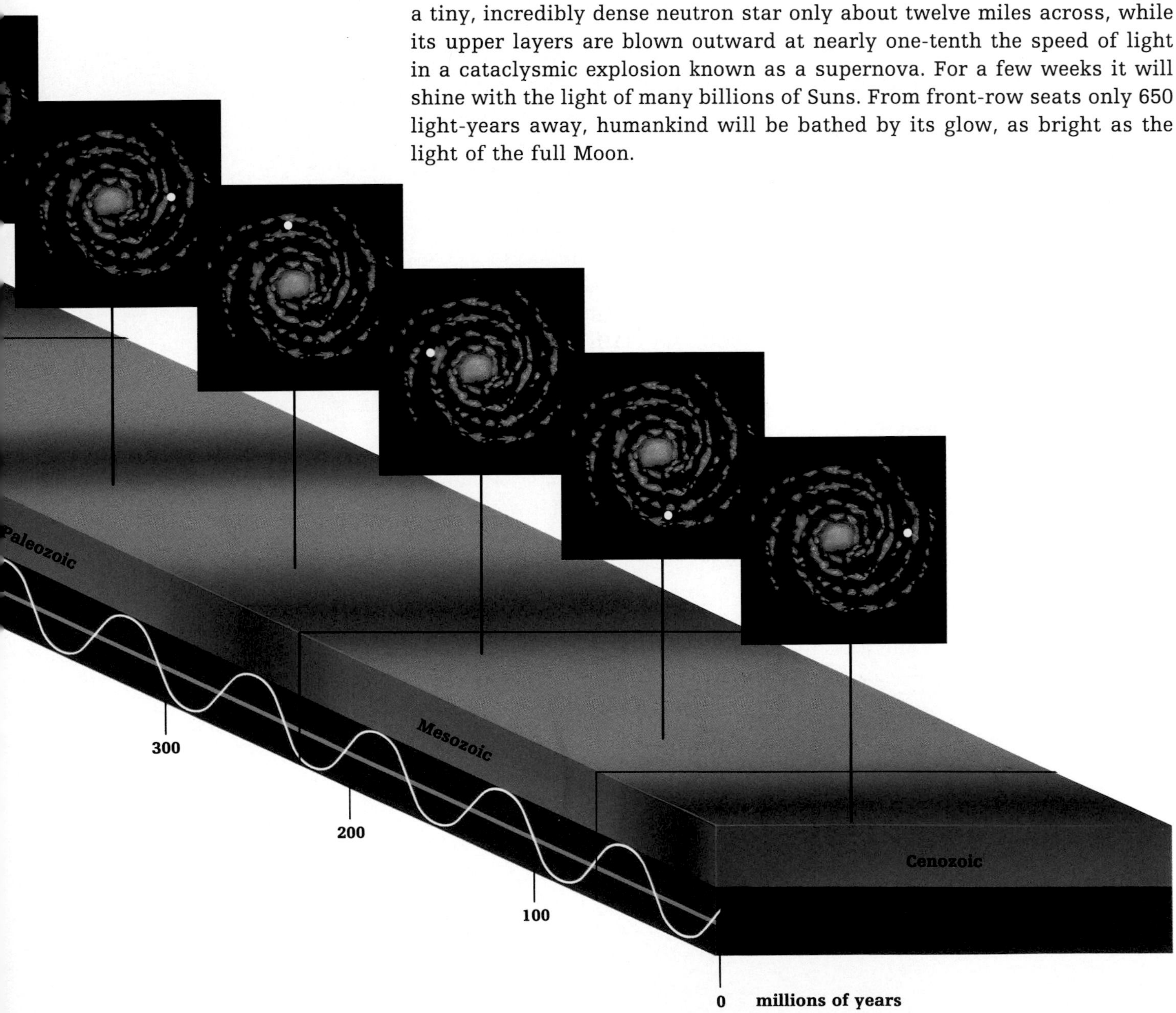

Paleozoic

Mesozoic

Cenozoic

300

200

100

0   millions of years

Supernovae have not been a common sight in recent centuries. The explosion named Supernova 1987A in February 1987 was the first reasonably close one since the invention of the telescope. But that rarity is probably a statistical fluke. Astronomers have observed many supernovae in other galaxies, leading them to estimate that one goes off somewhere in the Milky Way every 50 to 100 years. (Most are likely hidden by interstellar gas and dust.) This frequency raises a compelling point. All the known galactic supernovae have been far away enough to pose no threat to life on Earth; even Betelgeuse is comfortably distant. But if a supernova went off within, say, thirty light-years, the total energy falling on Earth's surface might be the equivalent of anywhere from 100,000 to a million hydrogen bombs.

For now, at least, Earth seems to be in no danger. There are only a few hundred stars within thirty light-years of the Sun, none with the requisite mass to make them supernova candidates. But just because there are currently no potential supernovae in the immediate vicinity does not mean that there will never be. As the Sun orbits the galaxy, it may one day encounter a star on the brink of detonation. Given the estimated number of supernovae in the Milky Way, a simple calculation indicates that one should go off within thirty light-years every 200 million years or so. Earth may have had to endure the horrific consequences many times in the 4.6 billion years since the Solar System was born—an event that itself may be traceable to a nearby supernova, whose shock wave could have instigated gravitational collapse in the dense interstellar cloud from which the Sun was fashioned.

Assuming there are still people around to watch the next time, a nearby supernova might be perceived as more exciting and beautiful than dangerous—at first. The initial harbinger, a burst of x-rays lasting about an hour, might go by unnoticed, since the radiation would be absorbed by the atmosphere. The next event, however, would be impossible to miss: As the x-rays began to fade, visible light and ultraviolet radiation, released slightly later by the explosion, would begin to increase. By the time they reached their peak intensity several days or a few weeks later, the supernova would be shining a thousand times brighter than the full Moon. For a brief spell it would seem almost like a second Sun.

Danger would come years or decades after that, with the arrival of a fierce blast of radioactive-like fallout in the form of cosmic rays—streams of highly energetic protons and heavier atomic nuclei with the power to damage living tissue and cause cellular mutations. Cosmic rays are nothing new, of course. Discovered early in the twentieth century, they steadily rain down on the planet from every direction, in meager-enough quantities that almost all are destroyed by collisions with atmospheric molecules. Although no specific sources have been identified, astronomers postulate that supernovae from hundreds of millions of years ago produced these particles, which have traveled through the galaxy ever since. The showers of cosmic rays from a nearby supernova, however, could be as much as a hundred times more intense. Furthermore, because the galaxy's own magnetic field would tend to deflect

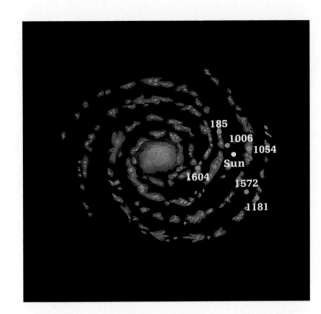

Astronomers estimate that supernovae occur somewhere in the Milky Way about once every 50 or 100 years, usually near the galactic plane, where the density of interstellar material tends to produce more massive stars. Because of the dustiness of the plane, earthbound observers have documented only six in the last twenty centuries, as shown above in a cross-sectional map of the galaxy. The explosions are visible from Earth only when they are relatively nearby or, as in the case of the supernova of 1604, when they happen above the plane, where there is less dust along the line of sight from Earth.

the particles and cause them to wander through space somewhat haphazardly, they would spread out and would keep coming for hundreds of years. Prolonged exposure to high doses would cause a proliferation of harmful mutations, perhaps to the point where many species would rapidly decline because few of their offspring would be viable. In order to survive, human beings might have to live underground, or establish colonies at the bottom of the oceans, beyond the reach of the deadly particles.

THE END OF A DELICATE BALANCE

As relentlessly threatening to life as these many forms of global turmoil and disruption have been and will continue to be, Earth's climate—though often imperiled—has proved astoundingly resilient, time and time again restoring conditions hospitable to life. The planet maintains a kind of natural thermostat known as the carbonate rock cycle that has been operating since the atmosphere first developed 3.6 billion years ago. Through a chemical process that geologists simply call weathering, carbon dioxide (CO_2), which is one of several so-called greenhouse gases in the atmosphere, combined with silicate rocks to form carbonate rocks (limestone and dolomites) that accumulated in layers some miles thick. Plants and animals assisted as well in preventing the buildup of atmospheric CO_2, locking up vast amounts of it in deposits of coal, oil, and natural gas.

But if the cycle prevented the atmosphere from becoming a deadly heat trap, it also maintained just enough carbon dioxide to keep the planet from freezing utterly. The carbonate rocks were constantly being pulled down into the mantle by plate tectonics and melted, whereupon their CO_2 was belched back into the atmosphere by volcanic action, setting the cycle in motion again. Geologists believe that this mechanism has kept Earth's temperature in balance over the eons. In a small way, ice ages fit into the pattern as well, their lower levels of precipitation allowing carbon dioxide to stay in the atmosphere and to reheat the globe.

Yet even this long-lasting temperature regulation is bound to fail. On the enormous time scale of the Sun's 10-billion-year life span, the delicate balance preserved through weathering grows increasingly fragile. Inevitably, of course, the Sun will go through its own death throes, dooming all life on Earth and ultimately rendering the planet and the rest of the Solar System unrecognizable *(pages 41-53)*. But climatic disaster lurks much sooner, well before the Sun becomes dramatically unstable.

Throughout its existence, as the Sun has burned its hydrogen fuel, the "ash" of that fusion reaction—that is, helium—has slowly accumulated in the core, where it contracts and heats up. In the process, the Sun has grown inexorably brighter. At its birth, it was roughly a third dimmer than it is today. Within the next billion years, it will brighten by 10 percent. Scientists estimate that at this point, the temperature-stabilizing mechanism will start to break down. Increased water evaporation from the oceans will have turned the whole world into a steaming jungle. The water vapor, which is itself a greenhouse gas, will be trapping more heat than ever. Clouds may slow the process by reflecting some of the sunlight back into space, but somewhere between 500 million and 1.5 billion years from now, Earth will reach a point when lakes, rivers, and oceans will simply dry up and disappear. And as volcanoes continue to release carbon dioxide, the greenhouse effect will intensify.

Earth's sister planet, Venus, offers a preview of what lies in store. Virtually the same size and of the same composition as Earth, Venus may very well have started out with oceans full of liquid water. But because it is closer to the Sun, the steadily increasing radiation of the evolving star caused its oceans to evaporate billions of years ago. As water molecules rose upward, the Sun's ultraviolet radiation split them into separate atoms of hydrogen (which escaped into space) and oxygen. Over the eons, carbon dioxide outgassed by volcanoes accumulated in the atmosphere to such an extent that the planet suffers surface temperatures of more than 1,000 degrees Fahrenheit—hot enough to melt lead—and surface pressures ninety-three times greater than the atmospheric pressure on Earth.

AND NOW, MARS

Even as Earth is transformed into a twin of Venus, however, its nearest neighbor, Mars, may undergo a temporary renaissance. Like Earth and Venus, Mars probably started out with a thick atmosphere of carbon dioxide and abundant liquid water; orbiting spacecraft can still see evidence for ancient water flows—runoff channels resembling dry riverbeds on Earth. But perhaps because the planet is smaller than Earth and cooled more rapidly after formation, plate tectonics appears to have ground to a halt early on in Martian history. Thus, with most of its volcanoes extinct and little carbon dioxide being recycled, the atmosphere was gradually depleted, until today there is just a thin wisp of it left. Mars's water is now locked away in frozen polar caps and vast layers of permafrost.

However, one billion years from now, at about the time Earth becomes completely uninhabitable, the Red Planet may once again support liquid water. In fact, by the time the Sun's luminosity has risen by 30 percent, roughly three billion years from now, temperatures on Mars could be approximately 45 degrees Fahrenheit warmer than currently. The ancient icecaps would melt and the permafrost thaw. Clouds and rainstorms would move through the newly moist atmosphere. The surface would become dappled by patches of iced-over water. The planet, though not exactly balmy, could feasibly make another home for humanity—at least for a while. For reasons having largely to do with its small size and resulting lack of tectonic activity, Mars is incapable of sustaining the carbonate rock cycle that maintains the atmosphere on Earth. All too quickly, the planet will lose its water vapor to space and once again be dry.

Regardless of what might befall terrestrial life in the distant reaches of the future, the fate of its home planet is grim. Five billion years from now, the Sun will have consumed the last of its hydrogen and grown into a red giant, with about 300 times its present luminosity. Earth will be a blasted hunk of rock, its deep oceans and thick polar ice long gone and its surface temperature approaching a staggering 1,400 degrees Fahrenheit. Then, about a billion years later, after a relatively brief shrinkage, the Sun will swell so enormously that its outer atmosphere might engulf the present orbit of Earth. But some astrophysicists think that by then—about 6.5 billion years from now—Earth itself will have moved.

Starting during its red giant phase and continuing even more vigorously during the next expansion phase, the Sun will be shedding a significant fraction of its substance in a prodigious outflow of gas and dust. Indeed, if that outflow resembles what astronomers have observed around other stars at this stage of their evolution, the night skies of the Solar System could be as bright as Earth's present daytime sky. In any case, the Sun could lose as much as half its mass in this way, reducing its gravity and allowing all the planets to spiral outward to twice their present orbits. Along the way, because the Sun will be 10,000 times more luminous than today, Earth's rocks will soften and melt as ancient continents slide into the empty basins that once held oceans. Eventually, the Solar System's former jewel will be covered by an ocean of molten rock.

One more transformation awaits humanity's home. Approximately seven billion years into the future, the Sun will undergo a final rupture, blowing off its distended outer layers into interstellar space. Astronomers have seen such expanding balls of glowing gas around other dying stars and dubbed them planetary nebulae because of their spherical, planetlike shape. At the center of each of these nebulae they also see the remains of the original star: a hot, superdense, but inert, lump of carbon and oxygen about the size of Earth. When the Sun reaches this so-called white dwarf stage, it will be less than a hundredth as bright as it is currently. With temperatures plunging below −325 degrees Fahrenheit—no warmer than Pluto is today—Earth's molten surface will freeze and turn glassy. And so things will remain for billions upon billions of years, the temperatures slowly dropping even further as the weak radiation from the ghostly white dwarf at the heart of the Solar System dwindles away to nothing.

SURVIVAL

In all of this hangs the question of humanity's fate. A few imaginative scientists and science writers have dared to speculate on the possibility of avoiding species extinction, more as an exercise in theoretical physics than as an attempt to predict the future. Astrophysicist Lloyd Motz of Columbia University, for example, has pointed out that with a few hundred million years of technological advancement, the heirs of humanity might be able to artificially regulate Earth's climate. The planet's inhabitants might manage this for several million years through such techniques as surrounding the globe with huge orbiting reflectors to bounce excess solar heat back into space. The reflectors might also include absorbing surfaces to trap some of the radiation and convert it to useful energy. Of course, notes Motz, when the Sun becomes a red giant, the time for technological fixes will have run out. Earth's passengers will then be left with only one possible strategy: abandon ship.

The approach has long been a staple of science-fiction stories that describe creatures from Earth advancing out into the Solar System's neighborhood as readily as Europeans migrated to North and South America in the centuries after 1492. Doing so would not only satisfy the deep-seated human urge to explore but, as the Russian space pioneer Konstantin Tsiolkovsky pointed out in 1911, would also increase the chances of survival: The more worlds the human race inhabits, the less chance that it will be wiped out by such random cosmic catastrophes as an asteroid or comet impact.

The new colonists' initial goal would be Mars, of course, and then the icy moons of Jupiter, Saturn, Uranus, and Neptune. But as one team of writers suggests, the most attractive homesteads may turn out to lie on the Solar System's myriad comets.

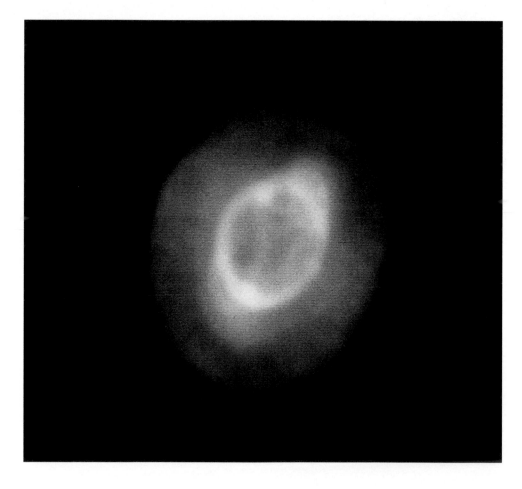

Some seven billion years in the future, the Sun will evolve into a planetary nebula similar to the one at left, NGC 3242, located about 2,500 light-years away in the constellation Hydra. As a low-mass star like the Sun nears the end of its life, it sloughs off an expanding, tenuous envelope of hot hydrogen that dissipates fairly quickly, in about 50,000 years. The remaining core will eventually collapse to form a dense white dwarf.

The idea may be a bit disconcerting at first. After all, as Carl Sagan and Ann Druyan admit in their book *Comet,* human beings evolved on a planet, and it is hard to imagine living on another type of body. But there may be trillions of comets out in the Oort cloud; taken together they have a surface area equivalent to 100 million planets the size of Earth. Moreover, even a modest-size comet, as little as a few hundred yards across, contains a million tons of material rich in the molecular necessities for life: water for drinking and for the extraction of oxygen; organic carbon compounds useful for agriculture and bioengineering; and probably significant amounts of rock and metal. For people with the technology to exploit these resources, comets might be far more hospitable than, say, the Moon, where water and carbon are in woefully short supply. And in the very long run, of course, comets would have the added advantage of orbiting so much farther from the expanding Sun than the Solar System's major bodies.

The only thing significantly missing in the cold domain of the comets in the outer Solar System is light. Equatorial noon at the distance of Saturn, about a billion miles from the Sun, is as dim as earthly twilight. And the fiercest sunshine out in the Oort cloud—trillions of miles away—might be little better than starlight. Still, colonists could always correct that deficiency by rigging vast solar collectors, or perhaps by building fusion reactors powered by deuterium, a heavy isotope of hydrogen that scientists have detected in the water ice of some comets.

One of the most intriguing, albeit bizarre, notions comes from Freeman Dyson, a professor of theoretical physics at the Institute for Advanced Study in Princeton, N.J. Dyson has long been noted as a keen-eyed visionary, and his most serious scientific efforts concern the distant cosmic future, but he occasionally engages in conjectures that might bear on humanity's fate. In the early 1970s, he gave some thought to comets as possible human habitats of the future and wrestled with some of their more significant deficiencies, namely the lack of warmth and air. The solution he proposed was typically whimsical: "We shall learn to grow trees on comets." In the low gravity, he pointed out, the trees' growth would not be restricted by weight: "From a comet of ten-mile diameter, trees can grow out for hundreds of miles, collecting the energy of sunlight from an area thousands of times as large as the area of the comet itself." The prospect was visually extraordinary: "Seen from far away," he wrote, "the comet will look like a small potato sprouting an immense growth of stems and foliage." The oxygen produced by photosynthesis would be transported down to the roots, and humans would live among the tree trunks. Humanity would therefore return to the arboreal life of its earliest ancestors. "We shall bring to the comets not only trees but a great variety of other flora and fauna to create for ourselves an environment as beautiful as ever existed on Earth," Dyson wrote. Going even further, he suggested that through biological engineering "we shall teach our plants to make seeds which will sail out across the ocean of space to propagate life upon comets still unvisited by man."

Taking Dyson's speculations further yet, Sagan and Druyan have envisioned a time when the Oort cloud would be dotted with millions of inhabited comets, each the dwelling place of no more than a few hundred individuals. These tiny colonies would presumably maintain contact with one another, but so great are the distances involved that even radio messages traveling at the speed of light could easily take a day or more to reach neighbors. As Sagan and Druyan conceive of this distant era, the cloud civilization would inevitably give rise to an enormous diversity of communities, each with its own social, political, economic, and religious views.

Moreover, as the billions of years go by, the cloud civilization might find the evolution and death of the Sun almost irrelevant. At that distance, the fierce light and heat of its expansive stages would simply make the gathering of solar energy easier for the comet dwellers, at least temporarily.

By that time, the colonists may well have taken the first steps toward seeding the rest of the galaxy with life. As they populate the Oort cloud, humanity's descendants will move outward through slow, gradual stages halfway to the Sun's nearest cousins—the three-star Alpha Centauri system, more than four light-years away. From there they might launch near-light-speed starships to the galaxy's farthest reaches, or even hook up engines to their tiny worlds and propel entire communities into interstellar space. Propulsion might not even be necessary: Individual comets would be so loosely bound by the Sun that a passing star could shake loose enormous numbers of them, and explorers could literally hitch a ride into the cosmos. Even if its destination is unknown, humanity's fate would seem to lie far beyond the realm of the Sun, among the stellar lights of the galaxy.

For just under five billion years, Earth's home star has flooded the Solar System with radiant energy, the product of fusion reactions deep in its core, where a vast store of hydrogen is steadily transformed into helium. The duration of this process, known as the main sequence phase of stellar evolution, depends directly on mass: The more massive the star, the more rapidly it uses up its hydrogen and the briefer its time on the main sequence. Stars with more than ten times the Sun's mass are the most profligate spenders of hydrogen. In less than half a billion years, they move off the main sequence and very soon thereafter explode as supernovae, blasting all interstellar space within a radius of about 300 light-years. Luckily for Earth and its planetary siblings, the Sun is an average-size star, whose finale will be much less sudden. The end of the Sun's main sequence phase is still some five billion years away, and no supernova cataclysm lies in wait beyond that. As shown on the following pages, however, the denizens of the Solar System can expect changes no less dramatic—and no less fatal—as the Sun negotiates the remaining stages of its long life.

In a sense, the changes have been happening all along. The present-day Sun shines some 40 percent brighter than it did in its infancy. As the newborn star grew in power, its burgeoning heat baked tiny Mercury and transformed Venus into a noxious oven. Many eons hence, those bodies will have vanished into the ravening depths of a Sun bloated and reddened beyond recognition, and the once-fruitful Earth will be barren. But farther out, the moons of Jupiter and Saturn will be emerging for the first time from their primordial deep freeze. A few billion years after that, however, the ice age will return for good. Whatever is left of the Solar System will circle the shrunken remains of the Sun, cooling inexorably to the frigid temperature of deep space.

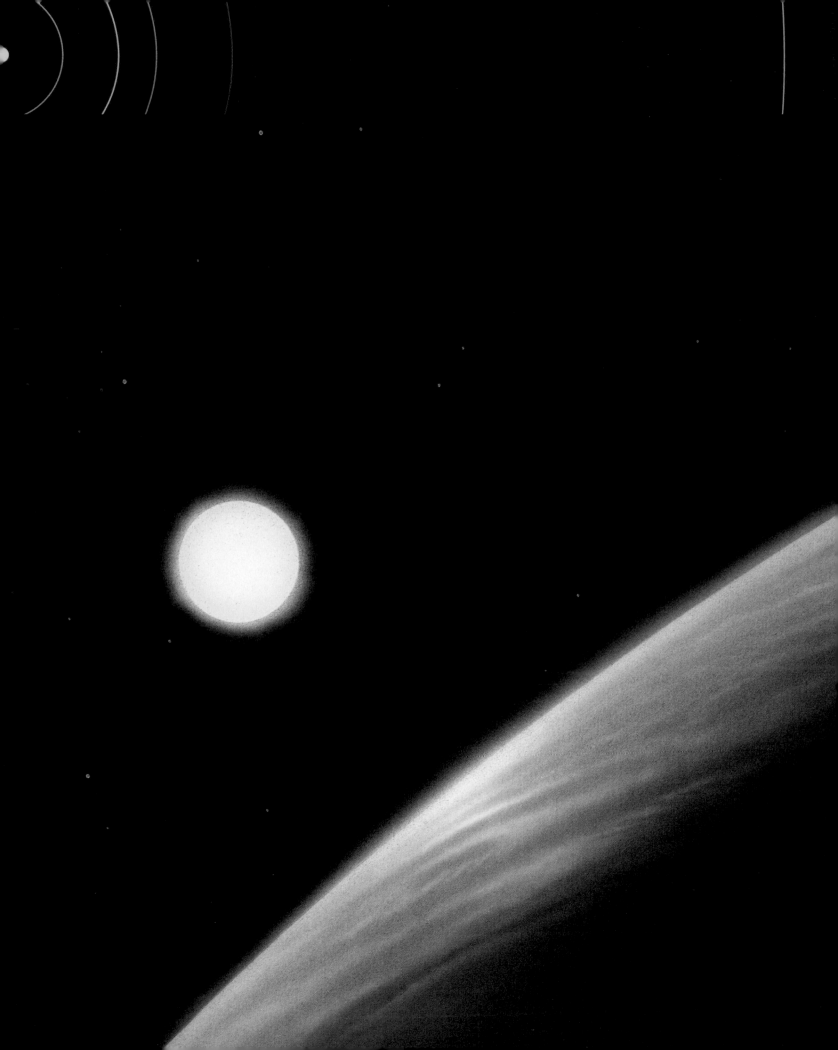

THE END OF LIFE
ON THE WATER PLANET

Over the course of the next several hundred million years, as the aging Sun grows ever brighter and hotter, the familiar blue-and-white globe that is Earth will undergo a terrible and irrevocable transformation. The intricate mechanism that has maintained the planet's temperate climate and life-giving oceans for some 3.6 billion years will suffer increasing strains, until at last it succumbs completely.

According to some calculations, the critical factor in the delicate balancing act is carbon dioxide (CO_2), one of several so-called greenhouse gases. When present in the atmosphere, these gases tend to let sunlight in while slowing the escape of heat reradiated from the surface, thereby creating an environment warm enough—but not too warm—for the existence of liquid water. Through the ages, Earth has been able to compensate for the gradual rise in solar heat by transferring CO_2 back and forth between the atmosphere and the crust, so that not too much of the gas accumulates in the atmosphere.

At present, very little of the carbon dioxide in the system resides in the atmosphere; most of it is locked in carbonate rock and is returned to the air slowly, through eons of geologic and tectonic action, only to be washed back down by rain. But roughly a billion years from now, the Sun will be at least 10 percent more luminous than it is today, and as the planets continue to hold to their current orbits *(inset, above)*, they will be subjected to a barrage of heat that will cause the cycle to break down. A complex of reactions will result in a buildup of atmospheric CO_2 and sulfates that will trigger what is known as a runaway greenhouse effect: Heat trapped at the surface will gradually dry up the world's oceans until Earth is blanketed by a stifling, sulfurous atmosphere *(left)* that leaves the planet as waterless as Venus.

A Temporary Respite on Mars

While Earth takes on a Venus-like aspect, Mars may enjoy a brief run as the new Earth. The increase in solar energy reaching the Red Planet will raise surface temperatures some forty-five degrees Fahrenheit, warm enough to thaw water ice and vaporize carbon dioxide that has been trapped in Martian topsoil since about 3.6 billion years ago, about a billion years after the planet's formation. Bodies of iced-over water might appear during summer in the middle to high latitudes *(left)*, and gaseous carbon dioxide would increase atmospheric density and pressure.

Even at the height of this Martian spring, however, with surface temperatures exceeding 100 degrees, Mars's atmospheric pressure will be scarcely one-third that on Earth. Thus, despite reserves of water ice plentiful enough to create a planetwide sea at least thirty feet deep, surface water will evaporate too quickly for such an ocean to form.

In any event, Mars's temperate phase will be short-lived. The planet, with only one-tenth the mass of Earth, is too small to have kept the internal heat needed to maintain geologic and tectonic activity. With no mechanism for recycling carbon dioxide washed out of

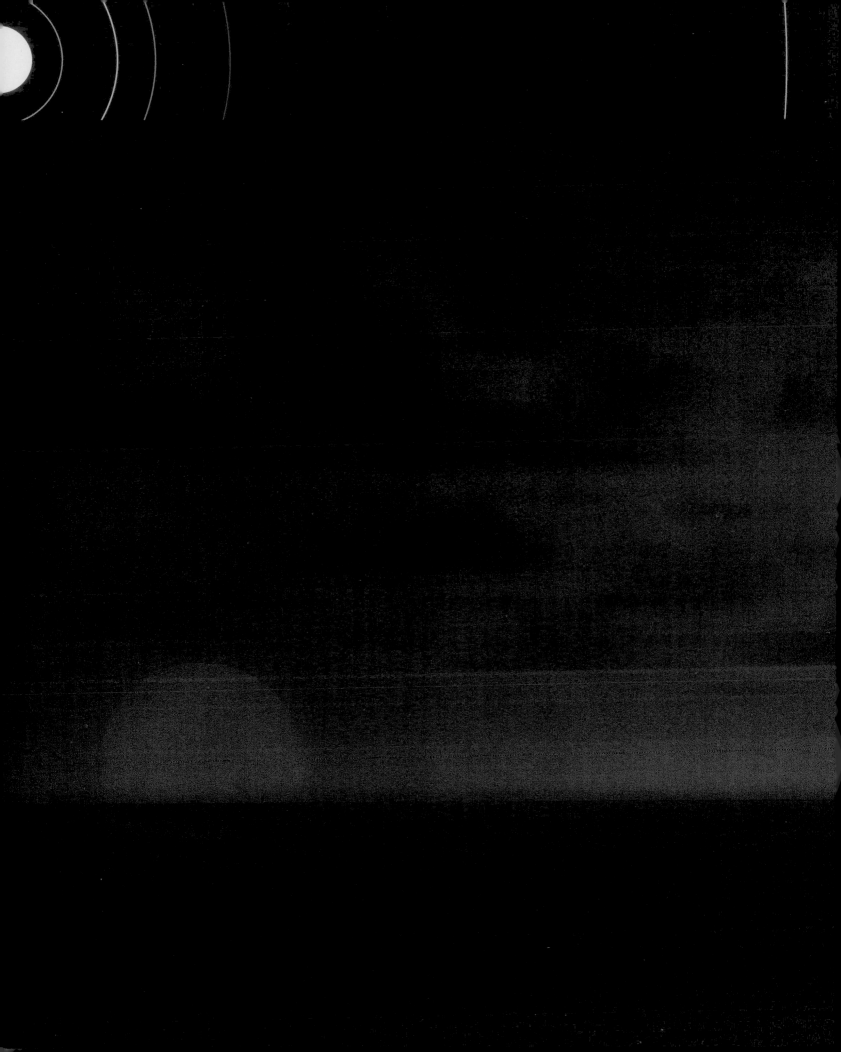

A Giant's Hot Breath on Icy Moons

Five billion years from now, the Sun will have exhausted its central store of hydrogen and entered the red giant phase of its evolution. As the mass of inert helium at the core compresses under its own weight, it will generate enough heat to ignite hydrogen in the surrounding layers and cause the Sun to balloon into a red giant forty times its present size. The gargantuan star will expand outward more than halfway to Mercury's orbit *(purple, above),* and although its surface will be relatively cool, its colossal size will make it an estimated 300 times brighter than today.

This period will have radical consequences for the outer planets and their attendant moons. Ice-covered Europa, Callisto, and Ganymede, for example, orbiting Jupiter 480,000,000 miles from the Sun *(orange, above)* and thus insulated by distance, will have kept their frozen sheeting throughout the Sun's main sequence crescendo of energy.

But when the red giant starts to grow, the moons will begin to melt. The swiftness of the thaw depends on the composition of the ice. If it is pure water, hundreds of millions of years will pass before it liquefies. A mixture of water and ammonia, however, will have a lower melting point, and the moons will be covered with water much earlier in the red giant phase.

Of all the Jovian moons, only Europa has a thick crust of nearly pure water ice. Data returned by the Voyager probes suggests that a liquid ocean eighteen miles deep may lie beneath this crust, prevented from freezing by the friction generated by Jupiter's massive gravitational pull. Once the crust has melted to expose this subterranean ocean, Europa will soon be shrouded in fog and clouds floating above an unbroken expanse of dark water *(left).*

TERRESTRIAL MELTDOWN

A billion or so years after oceans appear on the moons of Jupiter, a brief internal explosion known as the helium flash will set the erstwhile red giant on a new evolutionary course. The flash will signal the beginning of helium fusing to carbon in the Sun's core as continued compression causes temperatures there to rise to more than a hundred million degrees Kelvin. When all of the helium in the core has turned to carbon, energy production will cease, and the core will contract again under its own gravitational force. The resulting heat will eventually ignite an adjacent layer of hydrogen, generating a burst of energy that will launch the Sun into what is known as a second ascension, swelling it to such enormous dimensions that it engulfs Mercury and Venus and threatens Earth.

In stellar evolution, the second ascension is marked by so-called thermal pulses, fuel ignitions that alternate between two concentric shells around the core, one of helium and the other of hydrogen. With each successive pulse, the Sun will brighten and balloon, even as it continuously belches gas and dust from its outer layers into space.

Some seven billion years from now, the Sun will produce two to three thousand times as much energy as it does today. But the wholesale expulsion of as much as half the star's mass will eventually halt the thermal pulses in addition to diminishing solar gravity and weakening the Sun's hold on its planets. Assuming Earth survives the onset of the second ascension, it will have drifted outward to a new orbit *(blue, above)*, perhaps twice as far from the Sun as its station today. But the planet would be a melancholy sight: The outline of every continent would have long since melted into a featureless expanse *(left)*.

TURBULENT SHELL, SHRUNKEN SUN

At the end of millions of years of shedding its own matter, the Sun will consist of a core of carbon and oxygen—the products of helium fusion—surrounded by a thin, low-mass envelope that contains the last remnants of its once-enormous store of hydrogen. Near the still-hot core, this hydrogen will burn at temperatures of millions of degrees. The heat and the resulting luminosity will be so fierce, according to one theory, that the Sun will expel solar gas that will travel outward in all directions at speeds of more than a thousand miles per second.

This high-velocity gas will rapidly catch up with and slam into the slower-moving matter previously shed by the Sun, producing a turbulent shell of shocked gas known as a planetary nebula *(left)*. Expanding at the rate of about twelve miles per second, the nebula will grow ever more tenuous. Astronomers have tracked nebulae long enough to conclude that their life span is relatively short. In as few as 50,000 years, the Sun's nebula may spread itself so thin that it will simply disappear.

By the Light of a White Dwarf

Even as the Sun casts off the gases that will form the planetary nebula, its core—reduced to carbon and oxygen—will have become a white dwarf, a star with roughly half the mass of the original Sun compacted into a sphere no larger than Earth. This tremendous density will generate a corresponding heat, with surface temperatures rising above 100,000 degrees Kelvin. But with thermonuclear reactions long since shut down, this heat will eventually, and inevitably, dissipate into space.

All the Sun's surviving planets will have retreated in response to the star's loss of mass *(inset, above)*, and from the glassy, frozen surface of Earth *(left)*, the white dwarf Sun would appear no larger than the Venus of today, although it would bathe the landscape in light as bright as that of a hundred full Moons.

How long the Sun will spend cooling down is uncertain. But if current theories of stellar evolution are correct, the Sun will ultimately wink out. Then, in darkness and profound cold, the planets will spend a near eternity in orbit around a black dwarf.

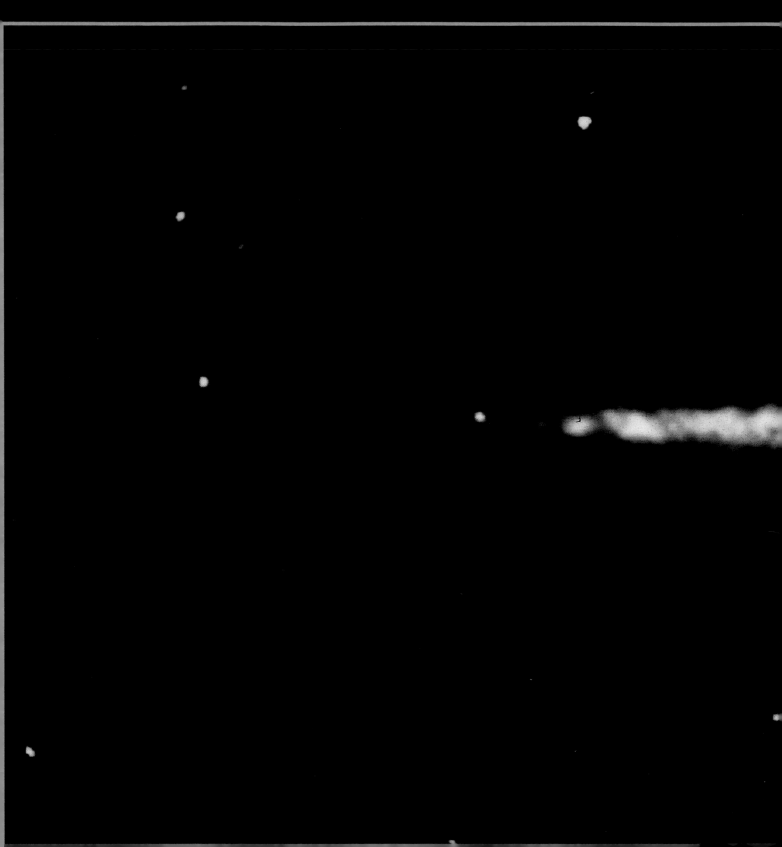

The radiance of billions of stars defines the central bulge of the Milky Way galaxy, which is located farther than 20,000 light-years away from Earth and is seen here in a composite infrared image. Despite its seemingly boundless energy, the Milky Way, like every other star system in the universe, is destined to expire.

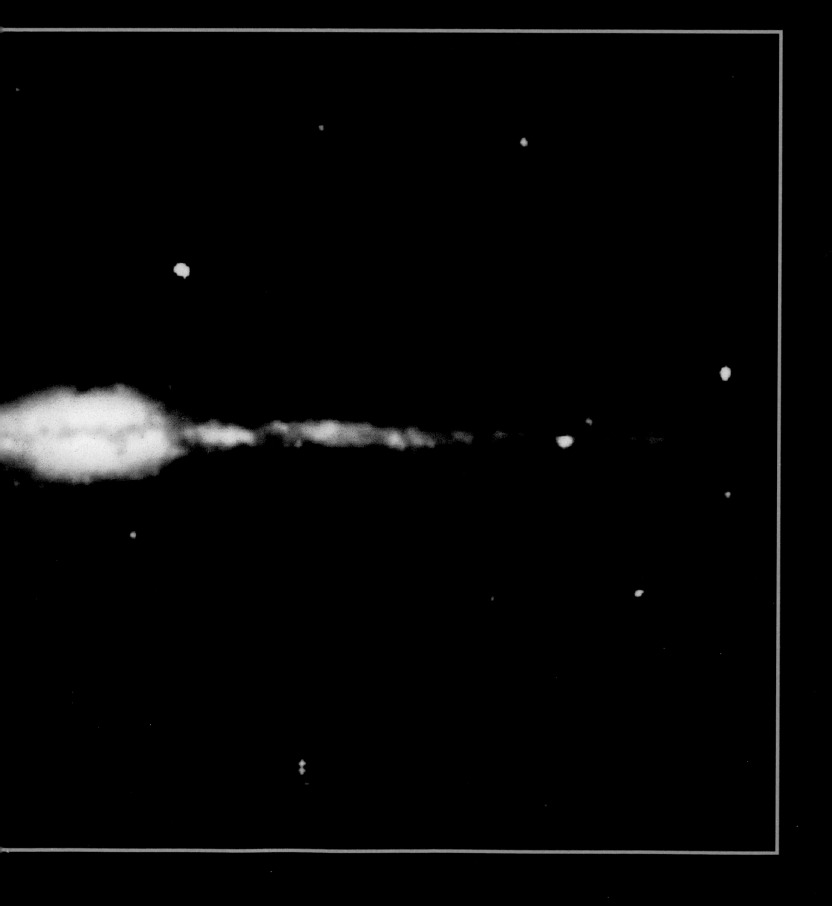

magine a time—15 or 20 billion years in the future—when the universe is twice as old as it is today. The once-vibrant Solar System is just a distant memory, at best a collection of frozen bodies orbiting the burned-out cinder of the Sun, which now has been dead for as long as it was alive. Most of the stars that populated Earth's skies in the twentieth century have also consumed the last of their thermonuclear fuel and passed into obscurity. If life from Earth has somehow managed to survive the turmoil of the Sun's death throes, it doubtless has taken up residence far afield, in the vicinity of some still-active star, whose light provides a steady flow of sustaining energy.

Yet for all the millions of stellar deaths that have taken place, the Milky Way galaxy is essentially the same as it ever was. Its center still harbors what today's astrophysicists assume to be a supermassive black hole, an object of unimaginable density whose crushing gravity prevents everything within several billion miles, even light itself, from escaping its hold. As matter dragged in toward the black hole spins wildly about the rim, the resulting friction and turbulence generate fiercely high temperatures and enormous quantities of electromagnetic energy that jet out along the hole's spin axis. Beyond the reach of this violent core and spanning more than 100,000 lightyears, the spiral arms of the galactic plane are still aglow with billions of stars interspersed among wispy clouds of dust and gas—the breeding ground for stellar generations yet to come. And all around the plane, a spherical halo of globular clusters—tight gatherings of anywhere from tens of thousands to millions of stars that follow eccentric orbits around the hub—defines the galaxy's outer limits.

Throughout the Milky Way and the many billions of other star systems inhabiting the universe, galactic business goes on as usual: Old stars die, new stars are born, and all manner of celestial objects keep up their orbital routines. But this sense of peaceful continuity is deceptive. Some galaxies suffer a tumultuous existence and meet untimely ends. Warped, torn, and twisted by close encounters and collisions with other galaxies, they can lose all semblance of organized structure and degenerate into indistinct associations of stars, dust, and gas.

Evidence is not hard to come by: The Milky Way's two nearest neighbors of the twentieth century, the Large and Small Magellanic Clouds, may be the shredded remnants of spiral companions that once were as elegantly configured as their more massive partner. By 20 billion years in the future, they may very well have disappeared altogether, consumed by the Milky Way.

The effects are not always so dire, and many systems survive such galactic run-ins, emerging changed but intact. In the final analysis, however, no galaxy is immortal. All have a natural, if incomprehensibly lengthy, life span that will eventually play out. The first major phase in a galaxy's end game involves the breakdown of its starmaking machinery, the chief element of which is the so-called interstellar medium, the dust and gas from which new stars coalesce. Galaxies have long depended on the stars themselves to re-stock this larder, but as time passes and increasing amounts of this fertile material get locked away in stellar corpses or expelled from the galaxy, the quantities available for the recycling process diminish. Ultimately, at a time dictated by the galaxy's supply of interstellar material, stellar genesis will grind to a halt, and as the last race of stars lives out its allotted existence, the galaxy will go dark.

For a very long time after their lights wink out, the galaxies themselves will persist: Planets will orbit stars, stars will orbit the galactic center. But finally even the most durable orbital relationships will begin to unravel. The dis-solution will involve the two essentially opposite processes of release and capture. In the vast majority of instances, orbital unions will be torn apart. Some dead stars, for example, will break free of their galactic homes to wander aimlessly, perhaps forever, in intergalactic space. In other cases, however, objects will no longer be able to keep their orbital distance, and larger celestial bodies will draw in their less massive partners and grow larger still. Ultimately, all that will be left of the galaxies will be gigantic black holes, bloated to enormous proportions by the vast quantities of mat-ter they have consumed.

Whether this grim scenario ever comes to pass, and what might happen next, depends on a related but separate issue—whether the universe as a whole will go on expanding forever or eventually collapse *(Chapter Three)*. But astronomers who would foretell galactic prospects must first understand how these stellar systems came into being and what mechanisms are at work in them today. The blueprints for their evolution were laid down shortly after the Big Bang—the fiery explosion that hurled time, energy, matter, and space violently into existence—in what most scientists assume were slight irreg-ularities in the dense stuff of the new universe.

EARLY CLUES
Building on the widely accepted model of the Big Bang, cosmologists have deduced that about ten seconds after the mammoth explosion, the entire universe—growing larger by the moment as space itself expanded—consisted of a superhot, seething amalgam of radiation and elementary particles. The temperature at this time was on the order of a few billion degrees Kelvin. (The Kelvin scale uses Celsius degrees and starts at absolute zero, equal to -273 degrees Celsius.) The radiation during these early instants was of such in-tensity that it completely dominated the particles, causing them to be so

energized that they moved too swiftly to come together and form atoms, remaining instead as random protons, neutrons, and electrons. These chaotic conditions prevailed for thousands of years as the universe continued to expand, all the time becoming cooler and less dense.

After several hundred thousand years, when the temperature reached approximately 3,000 degrees Kelvin, things changed dramatically. Radiation was no longer dominant, and electrons had slowed down enough to join with protons and neutrons—which had already united—to form stable atoms of hydrogen and helium, the two most common elements in embryonic space. As a result, photons of radiation that had been scattered by roaming electrons, creating an opaque glow throughout the universe, could propagate freely through space, which in effect became transparent.

Scientists call this the decoupling era, because radiation and matter became distinctly separate entities for the first time. The process left one of nature's fundamental forces, gravity, as the prime influence on matter and set the stage for the construction of organized units much larger than mere atoms, namely the galaxies.

Many cosmologists think that gravity needed a little help, however, for if the Big Bang had exploded uniformly in all directions, the early universe would have been perfectly smooth, and gravity would have had no reason, as it were, to exert more pull in any one location. Some scientists thus have suggested that prior to the decoupling era, the early universe must have contained so-called density fluctuations, minor variations in the concentration of matter from region to region. Although they cannot be sure what produced these fluctuations, adherents of this theory postulate that such an initial lumpiness enabled primordial gases eventually to condense under gravity's influence into enormous clouds. These clouds in turn would later fragment into amorphous masses known as protogalaxies—starless precursors of today's stellar systems.

According to this view, condensation began sometime between 200 million years and a billion years after the Big Bang. Then, as internal gravitational forces continued to cause gases in the protogalaxies to condense, the first generation of stars burst into life, within a span of about two or three billion years. Space kept on expanding, but galaxies that had originated in the same primeval clump of gas did not move away from each other; rather, they remained bound by mutual gravitational attraction, forming the groupings known as galaxy clusters that are in evidence throughout the heavens today.

Although this description apparently jibes with much of what is known about the current structure and distribution of galaxies, certain parts of the picture do not fit. For example, if there were density fluctuations before the decoupling era, they ought to have affected radiation as well as matter in the nascent universe. But researchers examining what is known as the cosmic background radiation—the cool afterglow of the Big Bang's initial emission, which has been diffusing through all of space ever since decoupling—can find no trace of unevenness. Nor is there yet a satisfactory explanation for how the

Astrophysicist J. Anthony Tyson of Bell Laboratories displays a liquid nitrogen container used to chill a light-sensitive electronic chip known as a charge-coupled device (CCD) to −148 degrees Fahrenheit. Attached to a telescope, the chilled CCD allowed Tyson and his colleague Pat Seitzer to register the whisper-faint light of galaxies so far distant that the radiation apparently began its journey nearly four billion years before the Sun and Earth were formed.

huge clouds of gas that preceded protogalaxies would have broken up, leading some investigators to argue that galaxies condensed out of the cosmic soup as completely independent units and only later, attracted by each other's gravitational force, came together to form clusters. Either way, a major challenge facing cosmologists at the end of the twentieth century is reconciling the evident smoothness and uniformity of the Big Bang with the undeniable tendency of matter to clump together on scales both large and small.

GALAXIES ABORNING

While theorists attempt to bring mathematics and physical principles into line with observations, some astronomers have taken a more direct approach to the question of galactic birth by attempting to actually witness the event as it happens—or, more precisely, as it happened. Because light has a finite speed of about 186,000 miles per second, any distant object appears to an observer as it was at some time in the past, when the light from the object began its journey to Earth. In other words, the deeper astronomers look into space, the farther back they are looking into cosmological history. With their most sensitive light-gathering instruments, scientists are attempting to capture the light emitted by newborn galaxies' first stars and by gas surrounding these stars. Perhaps this radiation will yield clues as to how the primeval gases broke up into protogalactic clouds and resolve the issue of whether all galaxies formed simultaneously, as current models suggest. Whatever the answers, they will have an important bearing on scientists' understanding of the future awaiting galaxies.

Searching for galaxies in the most distant reaches of time and space is a formidable endeavor, requiring sophisticated equipment and image-processing techniques, and sometimes great personal effort on the part of observers as well. The anticipated targets are so incredibly faint—some more than a billion times dimmer than celestial objects just barely visible to the naked eye—that exposure times are typically as long as six hours, and instruments can be focused only on portions of the sky devoid of bright stars and galaxies, whose luminosity would overwhelm the weak light of more remote objects. To bag their shy quarry, ground-based astronomers need the most nearly optimum viewing conditions possible, which often entails travel to remote mountaintop telescopes in such locations as the Chilean Andes, where the air is thin, clear, and dry, reducing the distortions caused by Earth's atmosphere to a bare minimum.

In 1983, astronomer J. Anthony Tyson *(page 59)* of AT&T Bell Laboratories in Murray Hill, New Jersey, and his colleague Pat Seitzer began a survey of twelve tiny patches of the night sky that previous studies had shown to be almost entirely empty. They used the four-meter telescope at the Cerro Tololo Inter-American Observatory in Chile and recorded their images with a charge-coupled device—an electronic detector that employs a grid of hundreds of thousands of light-sensitive cells to detect and count individual photons. By 1987, they had discovered about 25,000 faint, fuzzy light sources, some of which are almost certainly among the most distant objects ever observed, lying as much as 10 billion light-years away.

Tyson had little doubt as to the identity of the faint candles: "We are seeing galaxies as they were during their youth, just a few billion years after the Big Bang." If similar-size fields in the rest of the sky contain comparable numbers of these indistinct objects, Tyson speculated, then there might be 20 billion in all—a sizable proportion of the galaxies in the observable universe.

Although the data collected by Tyson and his colleagues has so far been little help in clarifying the process of galaxy formation, it does seem to indicate that one feature of the accepted view—namely, that all galaxies came to life during one splendid epoch of starmaking—may be flawed. The predominance of blue light from these objects suggests that a substantial fraction of the radiation is coming from very young stars and thus that these galaxies have been caught at an early point in their evolution. However, they also appear at a wide range of distances, the brightest ones from five billion to eight billion light-years away and others even more remote.

The obvious conclusion—in an expanding universe where differences in distance correspond to differences in time—is that galaxies did not in fact become active over a relatively short

A swarm of hundreds of galaxies fills the field of view in an optical image captured by a highly sensitive charge-coupled device. Spectral tests indicate not only that the star systems are as much as 10 billion light-years away but also that they are in the early stages of development. One surprising finding is that some of the young galaxies are several billion light-years nearer to Earth, meaning that they formed at a later time—contravening the theory of a single burst of galaxy formation a few billion years after the Big Bang.

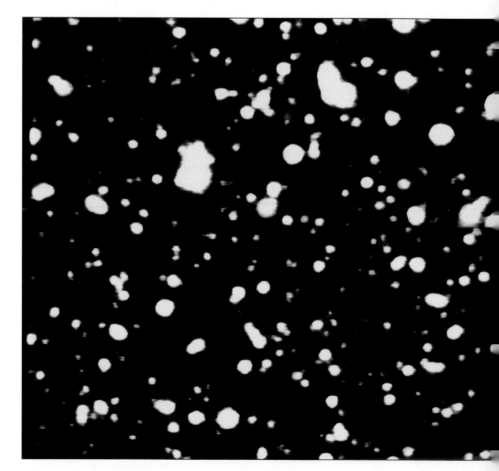

period but evolved gradually over several billion years. Although some scientists were initially skeptical about the accuracy of Tyson's distance measurements, the majority of the astronomical community now subscribes to this scenario. The most surprising aspect of Tyson's discovery, though, is how quickly after the Big Bang stars seem to have started forming. From what scientists currently understand about the mechanisms of gravitational collapse, nebulous gases should have taken much longer than a few billion years to clump together into stellar bodies. As Tyson put it, "I think these observations are beginning to constrain the theories. We may in fact come up with a major conflict that will give rise to a completely new paradigm."

THE SHAPES OF THINGS TO COME

Regardless of how galaxies got started, they have evolved into a diverse breed, with certain characteristic behaviors. Many of their traits seem to derive from their initial allotment of primordial material and from relationships established in the early days of the cosmos. Both influences will continue to be significant for long into the future.

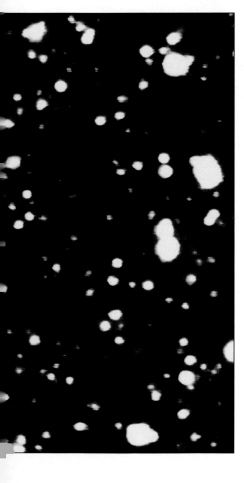

According to one view of galactic development, the nascent systems took on one of two distinctive shapes as they began to mature. In those whose gas was relatively clumpy, stars formed rapidly, using up the bulk of the galaxies' interstellar medium early on. These systems became the oval-shaped elliptical galaxies, almost entirely devoid of dust and gas and consisting mainly of older, reddish stars. Other infant galaxies, whose gas may have been more diffuse, formed stars more slowly and thus did not consume all of their stellar material in their youth. They also appear to have been subjected to more complex gravitational forces, both internal and external, which set them spinning more rapidly than the ellipticals and caused them to flatten out into disks. These went on to become the magnificent spiral galaxies, such as the Milky Way and its sister Andromeda. Still thick with interstellar dust and gas, they glow a bright bluish white with generations of young stars.

Ordinary spirals and ellipticals, though accounting for a good 90 percent of the galactic mass in the cosmos, are by no means the only members of the family, however. Among the most intriguing are AGNs, for "active galactic nucleus," a class that includes both spiral and elliptical systems. These galaxies are characterized by extremely energetic emissions from their cores, a degree of activity not associated with the normal stellar evolution that usually accounts for the light from galaxies. Among the first to be recognized as such were the so-called Seyfert galaxies, named after American astronomer Carl Seyfert, who discovered them in the 1940s. Seyferts are spirals with very bright, starlike nuclei that emit a broad range of electromagnetic radiation, from radio waves to x-rays. Later surveys identified a slightly dimmer class, dubbed N-galaxies (again, for the peculiar antics of their nuclei), whose brightness fluctuates, sometimes over the course of several months. Many of the N-galaxies are ellipticals.

In hunting for the evolutionary key to these energetic behaviors, scientists have focused on one of the most bizarre and enigmatic types of celestial object ever discovered—quasars. Discovered in the early 1960s, quasars got their name (a shortened version of "quasi-stellar radio sources") because they were initially revealed by radio-wavelength emissions that came from pointlike sources, as would light from a star. Astronomers have since discovered that only about 10 percent of quasars are strong radio sources, but all are compact emitters of intense x-radiation and characteristically have the extreme red-shifts that bespeak tremendous distance. The first quasars identified, for example, proved to be about three billion light-years away, and the most recent discoveries are located even deeper in space than the youthful systems detected by Anthony Tyson.

Many theorists have come to believe that quasars are the shining hearts of very young galaxies, the rest of whose light is, in effect, washed out by the brilliance of their nuclei. One plausible explanation for such incredible power is the matter-gobbling, energy-emitting behavior of a black hole, possibly similar to the one now thought to lie at the center of the Milky Way.

According to this hypothesis, an early galaxy that contained an especially thick concentration of primordial gas might generate enough gravitational force at its center to crush perhaps millions of Suns' worth of matter into a dimensionless point, thus producing a supermassive black hole. Since the galaxy would likely still contain an abundance of unprocessed gas, a rapidly spinning accretion disk would form around the hole, spewing forth enormous quantities of radiant energy that would be perceived billions of years later as quasar light. Relatively quickly, as matter was continuously sucked down into the hole, the accretion disk would be depleted, and the quasar light would fade, perhaps to the equivalent of the radiation from Seyferts and other active galaxies. Indeed, some scientists have suggested that the Milky Way itself was once a Seyfert and that its central black hole now has much less dust and gas to consume. Such a core may remain comparatively undemanding during the bulk of a galaxy's lifetime, but its unfading gravitational might is sure to play a major role in the system's latter days.

WHEN GALAXIES MEET

Galactic form, structure, and evolution, although often the result of internal dynamics, are also frequently affected by the fact that galaxies, much like the stars within them, are part of gravitationally bound systems. Astronomers now suspect that perhaps 70 percent of all galaxies are members of clusters that can contain only a few or as many as several thousand individuals. The Milky Way, for example, resides in the so-called Local Group, which includes not only the Large and Small Magellanic Clouds and Andromeda but some thirty other galaxies within a radius of approximately three million light-years *(page 67)*. Clusters themselves inhabit even more gigantic organizations known as superclusters, some of which stretch across hundreds of millions of light-years. Many observers believe that no galaxy is truly solitary, regardless of how locally isolated it appears

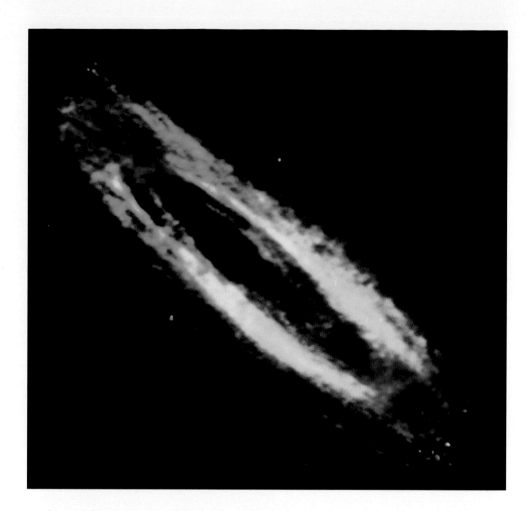

The Great Spiral in Andromeda, a large galaxy some two million light-years from the Milky Way, rotates counterclockwise about its axis as seen in this false-color radio image of hydrogen in the galaxy's spiral arms. (Blue signifies regions approaching along the line of sight from Earth; red marks regions moving away.) As it spins, Andromeda is inching toward the Milky Way, raising the possibility of a collision in the distant future.

to be; rather, it can always be counted as a member of a supercluster.

Often the distances between residents of a cluster or supercluster are so great that there is virtually no chance of their having any kind of influence on one another. In many cases, however, galaxies are surprisingly near neighbors. The average distance between bright galaxies is about 5 or 10 million light-years, yet in relation to their size—anywhere from several thousand to a few hundred thousand light-years in diameter—they are actually much closer together than are individual stars in the Sun's neighborhood. And, as the Local Group shows, many clusters are grouped even more tightly. In compact congregations, individual galaxies orbiting around the cluster's common center of mass occasionally come quite close to each other and can even collide. The Milky Way and Andromeda, currently separated by about 2 million light-years, seem to be in just such a predicament. Since the moment they came into being, their mutual gravitational attraction has drawn them together at the rate of about thirty miles per second, and even though this is a snail's pace in cosmic terms, within 10 billion years they should meet. Similar encounters that have already taken place may explain many of a large classification of galaxies, known as irregulars, that fit into neither the spiral nor the elliptical category.

The first detailed study indicating that galaxies may interact and thereby influence one another's shape was the work of astronomer Fritz Zwicky of the California Institute of Technology. Surveying the skies through the 200-inch Hale Telescope on Palomar Mountain in the 1950s, Zwicky found that a

A false-color optical image of galaxy NGC 1097 reveals intense energy emanating from its core, which may harbor a black hole whose terrific gravity is drawing in surrounding matter and generating vast amounts of radiation in the process. Similar emissions have been detected near the center of the Milky Way, suggesting that the galaxy contains a black hole of its own that could one day swallow the last remaining stars in the system.

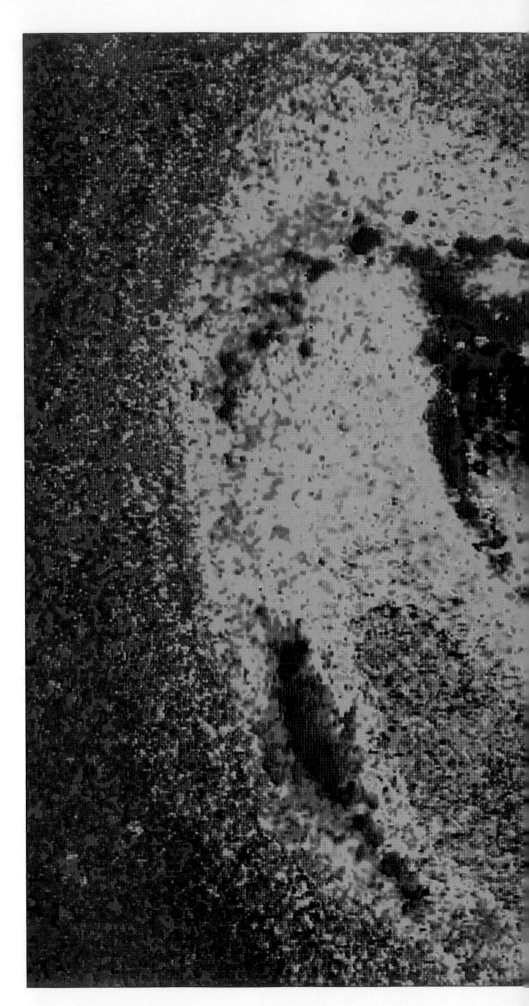

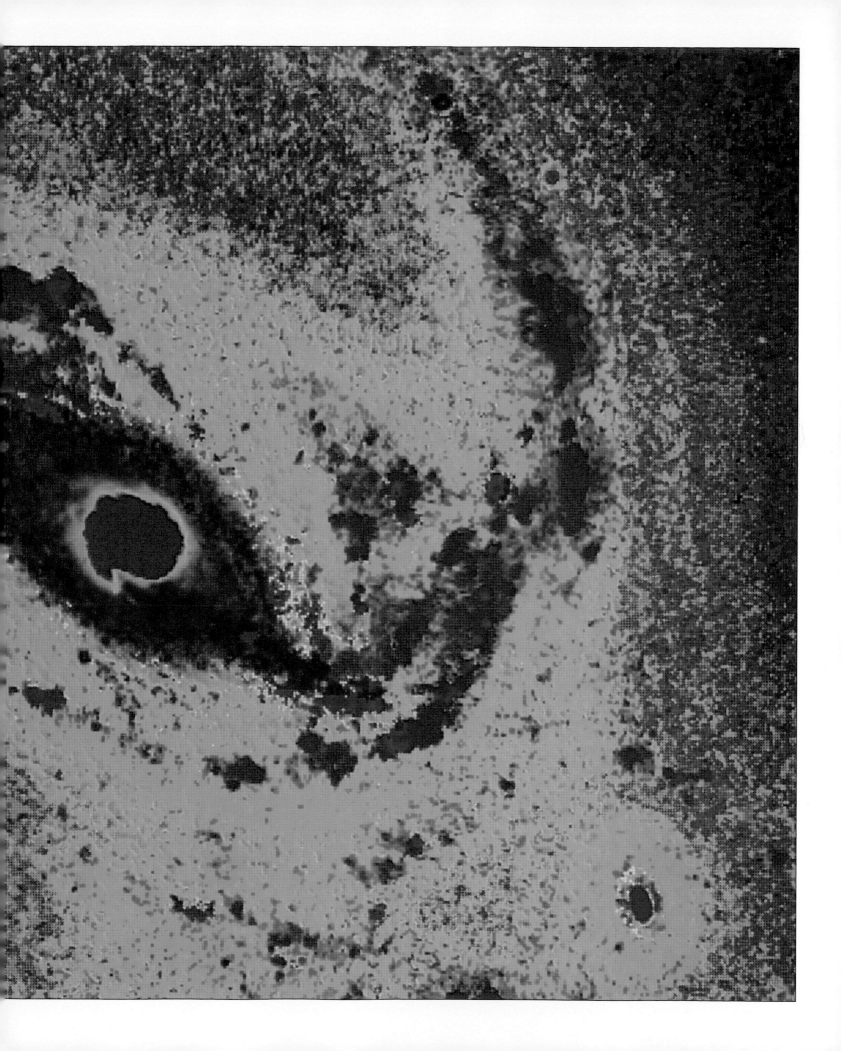

surprisingly large number of star systems appear to be connected by luminous intergalactic formations, and that many other pairs and groups of galaxies exhibit filaments of stars and gas streaming away from the galaxies' bodies. Zwicky suggested that the strange appendages and galactic bridges represented the aftereffects of gravitational forces exerted during close encounters, although he did not construct any theoretical models to support these conclusions.

Most astronomers of the day, convinced that galaxies were too widely separated for such behavior, dismissed Zwicky's examples as mere coincidence—an optical illusion created by the random placement of galaxies on the sky—and the theory languished for over a decade. But as observers began accumulating evidence of the clustering of galaxies and finding additional examples of apparent physical links between systems, increasing numbers of astronomers acknowledged that all sorts of interactions, from near misses to outright collisions, likely played a crucial role in galactic evolution.

Finding a way to test the theory is problematic. The scales of distance—and therefore time—are so great that observing a galactic encounter from beginning to end is impossible. Scientists have therefore sought confirmation from computer models that can reproduce events faster than they would occur in reality, simulating the velocities, masses, and relative orientation of galaxies on both flyby and collision courses.

In pioneering computer experiments at the Goddard Institute for Space Studies in New York in the 1970s, Alar Toomre of the Massachusetts Institute of Technology and his brother Juri Toomre of New York University (now at the University of Colorado at Boulder) set up a series of interactions between paired model galaxies. The staged run-ins generated all sorts of structures, from simple tails to skewed rings of stars and gas startlingly similar to features seen in photographs of real galaxies. Since then, improved processing techniques and faster, more powerful machines have enabled researchers to simulate a wide range of galactic encounters; the results have matched observations in enough detail to provide additional support for the theory.

A Group Destined for Encounters

Galaxies, like stars, tend to congregate. Just as a galaxy is defined by the stars in its gravitational thrall, the large-scale structure of the cosmos may be seen in associations of galaxies. As shown in the three-dimensional map at right, the Milky Way belongs to a cluster of galaxies known as the Local Group, whose members number about thirty. Since the distance to the various galaxies is calculated from Earth (the blue lines represent intervals of 650,000 light-years), the Milky Way occupies the center of the chart. But in fact, only a few galaxies, including the Large and Small Magellanic Clouds, orbit the Milky Way. The gravitational center of the Local Group lies somewhere between the Milky Way and Andromeda (M31), which has satellites of its own. Although the orbital characteristics of the group as a whole have yet to be defined, over the eons, momentum and gravity will inevitably draw some members into wrenching encounters.

The Local Group: (1) Milky Way.
(2) Draco. (3) Ursa Minor. (4) Small
Magellanic Cloud. (5) Large Magellanic Cloud. (6) Carina. (7) Sextans
C. (8) Ursa Major. (9) Pegasus.
(10) Sculptor. (11) Fornax. (12) Leo I.
(13) Leo II. (14) Maffei I. (15) NGC
185. (16) NGC 147. (17) NGC 205.
(18) M32. (19) Andromeda I. (20) Andromeda III. (21) Andromeda (M31).
(22) M33. (23) LGS 3. (24) IC 1613.
(25) NGC 6822. (26) Sextans A.
(27) Leo A. (28) IC 10. (29) DDO 210.
(30) Wolf-Lundmark-Melotte.
(31) IC 5152.

Powerful tidal interaction between the large spiral galaxy M81 *(left)* and the smaller, irregular-shaped NGC 3077 *(far left)* creates diffuse bridges of hydrogen gas between the two. (In this radio image, the highest concentrations of hydrogen are coded yellow.) Astronomers predict that M81 will ultimately draw most of the gas away from its neighbor into intergalactic space.

FATEFUL RUN-INS

Whether viewed on a computer screen or through a telescope, galactic interactions do more than just create convoluted shapes; they may account for a variety of effects that both enhance and hinder the standard evolutionary process. For example, disturbances of a galaxy's stable gas configurations or the transfer of gas between galaxies may be responsible for so-called starburst episodes, in which many stars suddenly come to life. According to this hypothesis, the collision or close passage of two systems can produce shocks that compress the galaxies' interstellar gas into dense clouds, which then collapse further under the drag of gravity to form new stars.

Cosmic confrontations can also be fraught with risk. Although the stars in galaxies are so far apart that few would actually strike each other even during a head-on galactic collision, the violent gravitational tides induced by the encounter can greatly influence the interstellar medium and the orbital dynamics of both stars and nebulae. Collisions between gas clouds thus may suddenly deplete galaxies of their starmaking material and even rip the gas and stars out of the galaxy. In fact, astronomers have identified a few galaxies whose unusual configurations suggest the disruptive effects of a recent collision. Among these are the ring galaxies, structures characterized by a bright circle of stars. It is believed that in at least some of these instances, the head-on impact of two galaxies is responsible. As one of the galaxies passed through the core of the other, its gravity generated a wave that traveled through the other's disk; radiating outward, the wave caused stars to form in an expanding loop.

Other phenomena result when interacting galaxies have very different masses. In these cases, the larger one may actually swallow the smaller in a form of galactic cannibalism. Moreover, in dense clusters, the devouring galaxy—having grown more gravitationally potent as a result of its meal—may continue to consume one galaxy after another to become a veritable

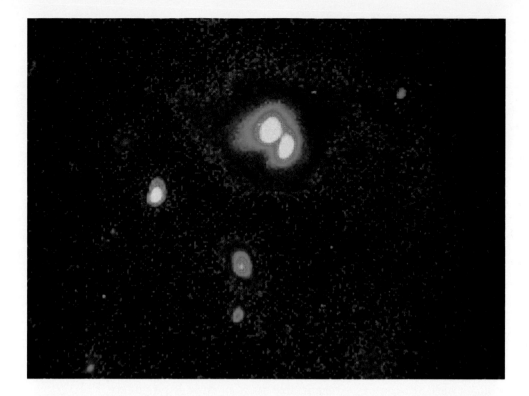

In another example of galactic interaction, two spiral star systems in the constellation Draco are merging to form NGC 6090, endowing it with double nuclei (coded white in this CCD view) and so-called tidal tails that are streaming out in opposite directions. The collision of gas clouds is promoting a frenzy of star formation around the double nuclei.

behemoth. Such an engorged galaxy exists in the cluster known as Abell 2199, located between the constellations Corona Borealis and Hercules; photographs of this giant's brilliant maw show what appear to be the remnant cores of three small galaxies. Because smaller galaxies greatly outnumber larger ones throughout the cosmos, some investigators postulate that every large system has had ample opportunity to capture many smaller ones. As evidence, they point to the rings of gas that orbit the rotational poles of some spirals, sitting like halos above the galaxies' central bulges; this gas may be all that is left of a loosely bound companion of the spiral.

A growing awareness of the role of mergers has led some galaxy experts to revise their thinking about one of the long-held assumptions concerning galactic formation. Rather than coming together in either disk or oval shapes, all young galaxies may have been disks, and ellipticals may actually represent a later stage in galactic evolution. Evidence indicates that elliptical galaxies predominate near the center of dense clusters and that they tend to be the largest members of such groupings. Astronomer François Schweizer of the Carnegie Institution of Washington is one of several scientists who have therefore proposed that ellipticals are the products of the collisions and mergers of spiral galaxies. Some computer simulations support the notion: Spirals in the process of interacting lose their disklike structure, converting most of their gas into new stars, ejecting some of their interstellar medium, and becoming gas-poor ovals. In addition, observers have detected several ellipticals that have some stars rotating in one direction and others in the opposite direction—further evidence that ellipticals result from the merger of two independent stellar systems.

Most astronomers agree that collisions and mergers were relatively frequent occurrences early in the universe's history, and that one way or another they have had significant influence on the development of galaxies. Furthermore, while these dire events may seem comparatively rare for the present, they are certain to have a profound effect on the galactic end game. Because clusters are likely to remain gravitationally bound as the universe expands, cataclysmic encounters will continue to occur, and the repercussions will accumulate. Over the vast stretches of time that lie ahead, monumental disruptions within clusters will take place, with disastrous consequences for the very existence of these galactic congregations.

THE STELLAR KEY

A number of the events that will affect the demise of the galaxies are already under way. At the most fundamental level, each galaxy's fate is linked to that of its stars, for when a galaxy ceases to blaze with stellar light, it is in a sense already dead. As stars expire, they return a portion of the matter that has been processed within them to space, where it is available for reincorporation into succeeding stellar generations. Just how efficient this recycling process is, and how long it can keep up the production of new stars, depends on the relative percentages of different-size stars making up a galaxy.

A star that has sloughed off its outer layers during its red giant phase and has a remaining core with a mass less than 1.4 times that of the Sun is destined to turn into a white dwarf and then, as it continues to radiate away its meager supply of energy, a black dwarf. Such low-mass stars return little material to

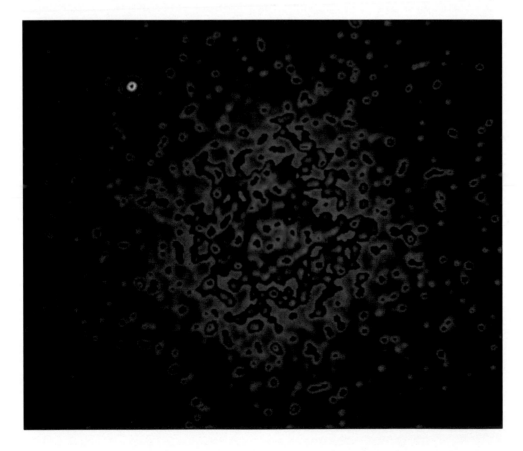

Scattered points of starlight surround a luminous core in this computerized image of globular cluster NGC 104—one of the round bundles of ancient stars that form a halo around the Milky Way. According to one theory, these relics of the primordial universe face "gravothermal catastrophe" as their dense core contracts, heating the cluster until stars at the periphery acquire so much kinetic energy they escape from the system—a process known as evaporation.

the interstellar medium as they dwindle away. Stars that contain greater than 1.4 solar masses in their core expire much more spectacularly. When they explode as supernovae, they radiate a hundred times more energy in a ten-second burst than the Sun will radiate in its entire lifetime, and they throw much larger quantities of matter back into space. Depending on the exploding star's mass, the remaining core may be crushed to a neutron star—an object considerably smaller than a white dwarf—or, if the core has on the order of, say, three Suns, to a black hole *(pages 72-73)*.

Although supernovae may provide enough matter to form some new stars, whether there are enough of them to significantly forestall the extinction of the galaxies seems doubtful. In the Milky Way, for instance, stars massive enough to go supernova make up a scant 4 percent of the galaxy's stars and contain only 11 percent of its total stellar mass. Many galaxies may be similarly proportioned. Ellipticals, for example, much like the globular clusters at the Milky Way's outer edges, tend to consist of less massive, slower-burning, and, hence, older bodies.

If galaxies are basically dependent on their original supply of gas, then, some may have enough to keep going for quite a long time, perhaps trillions of years. But theoretical physicist Jamal Islam, formerly of the University of Cambridge in England and now of the University of Chittagong in Bangladesh, has made a specific estimate that the Milky Way, for one, is comparatively close to forming its last batch of stars, which would reach their final stages in about 100 billion years. By the time a trillion years have passed, most galaxies will be strewn with the cooling corpses of dim stars. A few new stars, born within any remaining patches of dense gas or in the wake of occasional supernovae, will go on aging and dying for trillions of years more.

Throughout the universe, collisions, mergers, and seemingly insignificant grazings will continue, each encounter draining individual galaxies of their last reserves of gas and dust, and pulling off many of their stars. But whatever gas and dust is expelled into intergalactic space will be too widely dispersed to condense into new stars and will simply continue to expand into the void. According to such farsighted scientists as Jamal Islam and Freeman Dyson, the epoch of starmaking will come to a complete halt by 100 trillion years from now.

Dyson, who was devouring books on Einsteinian theory while still a teen-ager in his native England, is one of a handful of investigators who have delved into the topic of the physical structure of the future cosmos, and although he has tended to focus his attention on the most remote eons of the universe, long after the demise of galaxies, he has had his share to say about events along the way that do involve star systems.

The work of theorists like Dyson and Islam suggests that the once bright and sparkling galaxies will consist of dead stars, black holes, and other lifeless bits of matter, such as frozen planets, asteroids, and vague wisps of dust and gas. Occasionally these systems, which will still be held together by gravity, may emit short bursts of electromagnetic radiation as traces of mat-

THE FATES OF STARS

Stars follow sundry paths in their journey from infancy to extinction. Some are destined to die violently after relatively brief and turbulent careers; others will simply fade away after shining steadily for eons. The critical factor that determines a star's fate is mass: The larger the star, the faster its evolution and the more spectacular its demise. As shown here, astronomers chart a star's future by gauging its mass in relation to the Sun and placing it within one of several categories—a ranking that foretells whether the star will dash toward an explosive doom or plod its way into posterity.

The gargantuan fusion furnace at the heart of a blue supergiant—the size of 100 Suns—will run through its supply of hydrogen in just a few million years. The star will then inflate to many times its original size (proportions here are not to scale) before going supernova. The star's heavy core, left behind, will collapse into an infinitely dense black hole *(left)*.

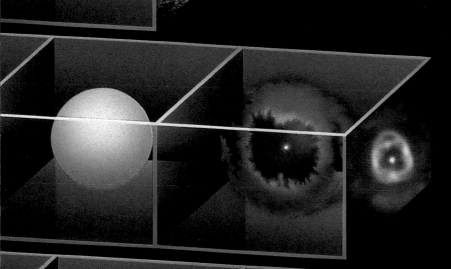

A star of ten solar masses will burn its fuel at a more temperate pace over tens of millions of years. But it too will eventually swell into a supergiant and go supernova. The surviving core lacks sufficient mass to produce a black hole. Instead, it shrivels into a neutron star, a body about twenty-five miles wide with the density of half a million Earths.

A star of roughly the Sun's girth—one to four solar masses—paces its consumption of hydrogen over 10 billion years or so, and its relatively stable core never reaches the explosive stage. When the hydrogen is nearly exhausted, the star balloons into a red giant a hundred times its original size. Then it sheds its gaseous envelope as a circular planetary nebula that drifts gradually away from the core, a dying ember known as a white dwarf.

The tiny red dwarf, just one-tenth the mass of the Sun, is the steadiest, most reliable furnace in the universe. With little gravity to accelerate the fusion in its core, the star will burn gently for several hundred billions of years. In the unchartable future, when its hydrogen reserves are finally depleted, the stellar Methuselah may fade into a black dwarf, a theoretical dark sphere of compressed gas.

ter are pulled into either stellar or supermassive central black holes. For all practical purposes, however, the brilliant heyday of the galaxies will be over.

FROM UNITY TO DISSOLUTION

Even though they will no longer shine, galaxies will still exist as organizational units, and the doomsday clock of the cosmos will continue to tick. In fact, for some hundred times as long as they glowed with active stars, the galaxies will carry on relatively unchanged, obscure inhabitants of a bleak and chilled universe. Some of the stellar corpses within galaxies will still have a gravitational hold on their long-dead planets, which will dutifully continue to orbit. But the unions will not last. Just as galaxies are bound to continue meeting or passing close by each other as time runs on, fast-moving star-planet systems will brush by other systems again and again. After about a hundred such close encounters, planets will be boosted by a kind of gravitational slingshot effect to a velocity sufficient to fling them out of orbit, free of their stars—in much the same way as probes from Earth pick up speed by swinging around other planets. Freeman Dyson has estimated that close encounters between planet-bearing stars should release most satellites within 100,000 trillion years.

Although planets will be first to feel the cumulative effects of these numerous passages, the dead stars themselves will also be disturbed. Intrigued by the possibilities, Jamal Islam in 1979 published a paper that explored, among other things, the ramifications of one type of encounter, between a binary pair and a third star. Islam worked out the details of what would happen when one member of the pair was less massive than both its partner and the third star. His calculations indicated that the smaller, and therefore faster-moving, star of the binary pair would pick up energy from the intruder and escape the clutches of its larger mate. If the encounter was particularly close, the fleeing star would probably achieve enough velocity to break free from the galaxy. Meanwhile, the intruder would probably lose enough kinetic energy to be captured by the remaining star, and because of their size, they would become bound in an even tighter gravitational relationship.

The same principles would apply to many different kinds of multiple-body encounters. Islam acknowledged that such meetings would be exceedingly rare—only a handful occurring every billion years or so—but given enough time, the results would be dramatic. Because of the indefinite probabilities involved, Islam would only make the broadest of estimates. His prognosis was that, anywhere from a million trillion to a thousand trillion trillion years from now, 99 percent of a typical galaxy's dead stars might escape in this way. The percentage might seem excessive, but it was in part based on the fact that most stars are on the low-mass end of the scale and at this point would have degenerated to the black dwarf stage. They would thus be catapulted out of their galaxies in encounters with the more massive stellar corpses of neutron stars and black holes.

Although the astronomical community is largely reluctant to engage in such

A Dynamo That
Twists Space and Time

As stellar evolution proceeds on its inevitable course, galaxies will become increasingly populated by the invisible enigmas known as black holes. Created from the collapsed cores of giant suns, black holes are unimaginably dense, their entire mass packed into an area known as a singularity. The resulting gravitational field is so powerful that it not only sharply warps but even opens a hole in the surrounding fabric of space-time—shown here as a two-dimensional sheet representing the universe's three dimensions of space and one of time.

Because most stars rotate, and the laws of physics dictate that a rotating object spins faster as it contracts, a black hole formed from a star is likely to be spinning at nearly the speed of light. The black hole's rapid rotation pulls space-time around with it, a whirlpool-like effect known as frame dragging (indicated here by the spiraling lines within the black hole's gravitational well). The following pages illustrate how a civilization of the distant future might use frame dragging to tap a spinning black hole's immense store of rotational energy, thereby sustaining life in a galaxy devoid of other energy sources.

Because physics cannot yet account for the infinitely dense mass of a singularity, the exact nature of this region remains a mystery. Within a certain radius of the singularity, gravity's pull is so intense that nothing, not even light can escape *(central black area at left)*. The outer limit of this zone is known as the event horizon. Beyond the event horizon is the ergosphere *(purple)*, a region where no object can resist the frame-dragging effect of the spinning singularity. Everything in the ergosphere, including space itself, is pulled around in the direction of the hole's rotation. Although nothing can stand still, an object moving close to the speed of light can escape from the ergosphere.

HARNESSING A COSMIC GENERATOR

In 1969, British mathematician Roger Penrose proposed a method for tapping the tremendous energy of spinning black holes. His elegant mechanism, subsequently dubbed the Penrose process, takes advantage of the complex physics at work in the ergosphere, the region where frame dragging is irresistible. If an object is sent into this zone and breaks in two so that one part falls into the black hole, the other part can be propelled outward with greater energy than it originally possessed. Penrose's calculations show that the extra energy of the escaping object comes from the black hole's rotation.

A few years later, American scientists Charles Misner, Kip Thorne, and John Wheeler expanded on Penrose's theory in a textbook on gravity. Their somewhat whimsical idea forms the basis for the scenario depicted at right. An advanced civilization has constructed an enclosed bubble five million miles in diameter around a black hole containing ten times the mass of the Sun and rotating in a counterclockwise direction at about 94 percent of light-speed. A space capsule hauling a garbage barge is launched at two-thirds the speed of light, fast enough to allow a close approach and escape. Whipped into an inward-spiraling orbit, the vehicle enters the ergosphere *(purple)* in the black hole's equatorial plane—where the ergosphere is widest—and at a given point, the capsule ejects the garbage barge backward with enormous force. The barge spirals down into the black hole while the capsule picks up rotational energy and flies back out to the bubble's inner surface, where it strikes a flywheel that powers an electrical generator. The process is so efficient that, under these conditions, two tons of garbage could yield the equivalent of Earth's total annual consumption of energy.

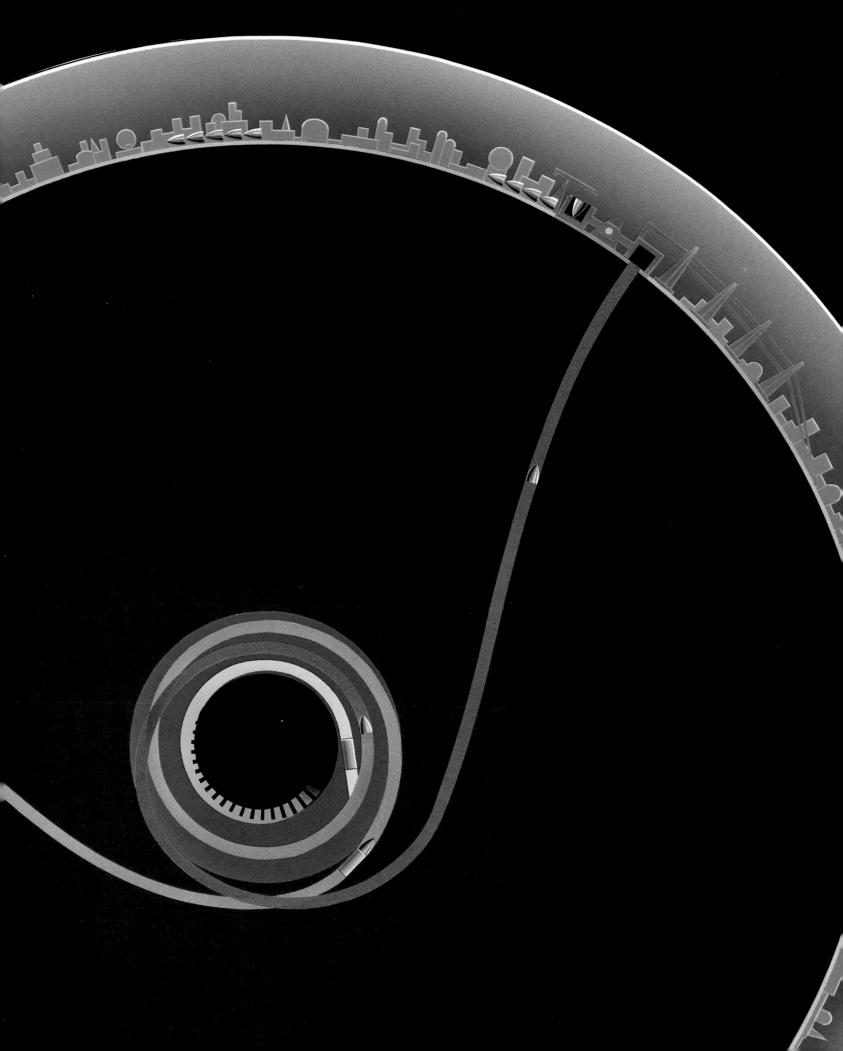

long-range speculation, a few scientists have followed up on Islam's predictions. Duane Dicus of the University of Texas at Austin and three of his physicist colleagues worked out the details anew in a 1983 paper and reached similar conclusions, calculating a 90 percent loss of mass by the galaxies over the course of a million trillion years. The authors characterized the process of star loss as galactic evaporation, because it would, in their words, "reproduce on a majestic scale the interactions of molecules evaporating at the surface of a liquid."

As for the neutron stars and black holes that are left behind, Islam demonstrated that the strong pull of their large masses would cause them to be gathered together in ever denser and more massive unions. And because of the loss of kinetic energy to the escaping stars, the remaining galactic inhabitants would become too weak to fight off the powerful gravitational force exerted by each galaxy's core.

At this point, it would make no difference whether a galaxy had started out with a black hole at its heart or not. As the entire galactic system spiraled in on itself, either its central black hole, trillions of years old, would swallow up whatever was left, or the steadily increasing accumulation of mass at the galaxy's hub would naturally produce such a locus of inconceivable density. All black holes are, of course, dimensionless, but they do exert an inescapable gravitational pull out to a clearly defined point in space known as the event

CLUSTER EVOLUTION

In a typical galactic cluster *(below)*, spirals dominate the fringes, and ellipticals prevail toward the center, which may be occupied by a giant elliptical. Galaxies orbit this core in various directions, setting the stage for epic encounters *(right)*. When galaxies merge, whatever their original shape, they tend to form ellipticals. The largest, including the giants at the hearts of clusters, may then act as cannibals, gorging on their smaller neighbors.

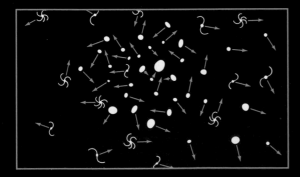

In Stephan's Quintet, which is situated in the constellation Pegasus, four of the five galaxies seem to be gravitationally ensnared and beginning to move toward one another.

horizon. A typical supermassive black hole containing, say, five million solar masses has an event horizon of about 9,200,000 miles—approximately a tenth the distance from the Sun to Earth. But a galactic black hole with the mass of a billion Suns, such as might result from the collapse of the Milky Way after 99 percent of its mass has evaporated, would dominate the space around it out to a radius of nearly two billion miles—the current distance of Uranus from the Sun. In galaxies up to 100 times more massive than the Milky Way, the ultimate black holes would squeeze together as much as 100 billion solar masses and have event horizons more than 50 times wider than Pluto's orbit.

THE END OF ORBITS

Freeman Dyson, anticipating every possible variation in the universal grand finale, has called attention to yet another dynamic mechanism that will become increasingly significant as the future unfolds and will work to destroy the last vestiges of gravitational relationships in the cosmos. One of the fundamental principles of relativity is that the universe is a continuum of the three spatial dimensions and the dimension of time. The continuum has often been visualized as a stiff but flexible sheet that is warped and dented by gravity; the more massive an object, the greater the degree to which it warps the sheet. Furthermore, any moving object—such as a planet orbiting a star, or a star moving through a galaxy—generates so-called gravity waves, ripples

The galaxies of a group known as VV172 are closing ranks, soon to merge into one large elliptical. (The object second from the right is not part of the gravitational system but lies much farther from Earth.)

The bloated form and wispy loops of the giant elliptical galaxy NGC 1316 indicate that it has recently consumed another galaxy; the disk-shaped system just above it may be next on the list.

in the space-time fabric that actually represent a slow depletion of the object's gravitational energy. The effect is so infinitesimal that it has never been detected and is of little consequence in the short run. As in many other instances, however, time will magnify even the slightest of effects. Applying this aspect of celestial dynamics to the end of galaxies, Dyson has speculated that all orbital relationships will eventually break down as gravity waves carry off kinetic energy from every object in revolution around another. Satellite bodies will be unable to maintain their orbits and will come crashing into their larger partners.

Dyson's figures indicate that the process of galactic evaporation will probably reduce galaxies to black holes before gravitational radiation will have a noticeable effect, but other associations may survive long enough to feel its influence. If, for instance, an entire planet-star system somehow escaped its galaxy intact, orbital erosion through the emission of gravity waves would ultimately finish the job. Using the example of Earth and the Sun, Dyson has calculated that the two would collide through gravitational radiation in 100 million trillion years.

Moreover, whatever conditions hold for planets and stars will hold for the gigantic black holes that are the monstrous remains of galaxies, some of which may still be gravitationally bound together in clusters. Whether through a process similar to the evaporation of stellar corpses from galaxies or through gravitational radiation, these last bonds will also be broken. At the heart of clusters, former galaxies will fuse into colossal supergalactic black holes ten to a thousand times more massive than single galactic black holes. The distances from one former cluster to another will preclude the chance of any further encounters.

Finally, then, perhaps as long as a thousand trillion trillion years from now, all celestial associations will have come to an end. Across the unimaginable vastness of the universe, a dreary collection of lightless objects—dead planets, random particles of dust and gas, black dwarfs, and black holes of all sizes—will wander space alone.

STARMAKERS NO MORE

From Earth's vantage point in the Milky Way, galaxies appear to be the building blocks of the cosmos. Everywhere astronomers look, vast spiral pinwheels and glowing ellipses group themselves into ever-larger aggregations that point to a universal coherence and structure. But this structure is of finite duration. Although the magnitude of change is still trifling—and will remain so for many trillions of years—galaxies are already at the mercy of forces that will result in their ultimate destruction.

The forces are twofold in nature. First, as the individual stars that make up a galaxy reach the end of their lives, they lose mass—whether abruptly in supernova explosions or gradually, sloughing off their outer atmospheres over time. In so doing, they increase their vulnerability to gravitational disruptions by passing stars. Even as these small-scale inner changes take place, entire galaxies are themselves subject to near misses and collisions of cosmic proportions, resulting in distortions that involve millions of stars as well as massive quantities of the interstellar gas and dust that give birth to new stars.

Astronomers attempting to describe the consequences of these complex phenomena often rely on sophisticated computer simulations, mathematical models that can be translated into graphic form. As shown on the following pages, illustrations based on such simulations as well as on numerical calculations represent scientists' best efforts to visualize the unfolding of countless eons of galactic evolution. Each stage depicted is accompanied by a time line that marks, in powers of ten, the approximate number of years between now and the onset of a given phase of the death of galaxies.

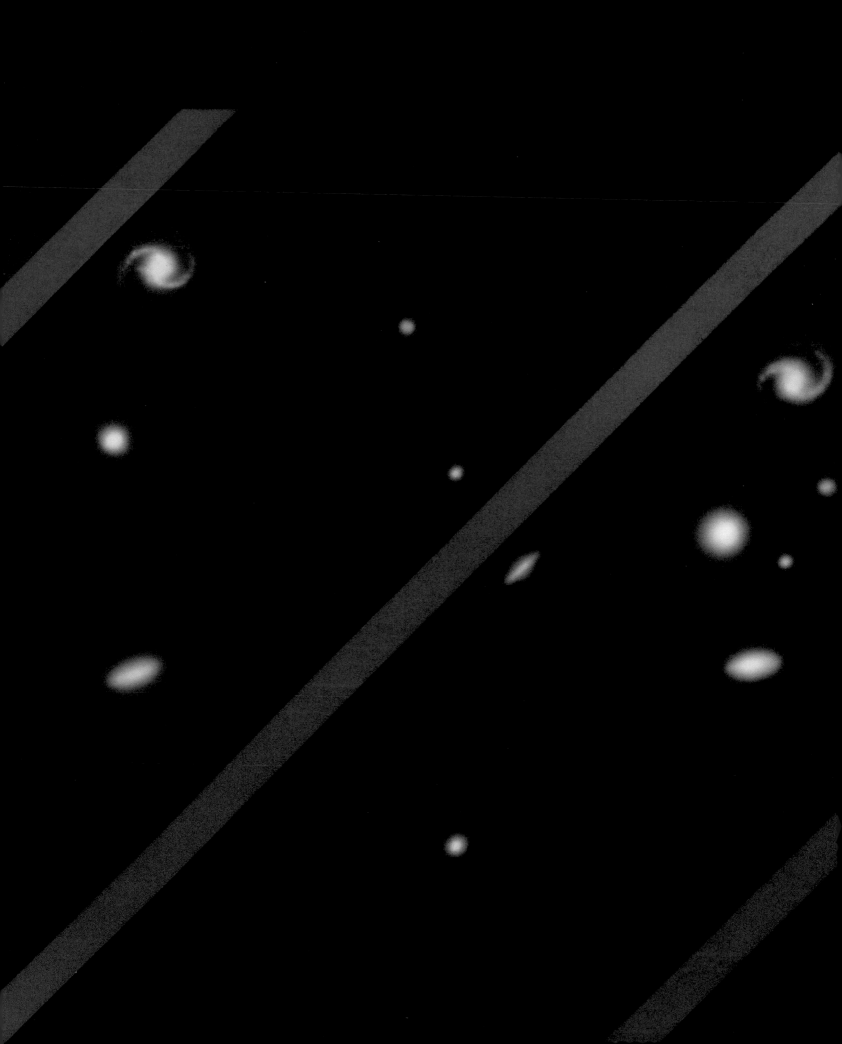

Cannibalism on a Grand Scale

In today's universe *(far left)*, some 15 to 20 billion years after the Big Bang, very few galaxies exist in isolation; most congregate in clusters that number from a few to several thousand members. And although galaxies come in many sizes, their shapes are remarkably similar. More than 60 percent of them are flattened disks similar to the Milky Way, a configuration that is believed to be the natural outcome of the gravitational collapse of the rotating nebulous clouds from which galaxies form.

As the members of a cluster orbit their common center of mass, some may draw near enough to disturb one another gravitationally, producing peculiar-looking systems with trailing appendages of stars and gas that have been described as tails or bridges. When such encounters are full-fledged collisions, however, the result is often a merger that may yield a single spherical or football-shaped elliptical galaxy.

Over time *(middle)*, gravitational attraction and the resulting energy exchanges will cause galaxies in clusters to converge, increasing the frequency of encounters and decreasing the number of disk galaxies as they merge into ever-larger ellipticals. A trillion (10^{12}) years hence, galaxy clusters may be reduced to giant ellipticals *(below)* that have collided with and swallowed most of their disk-shaped neighbors.

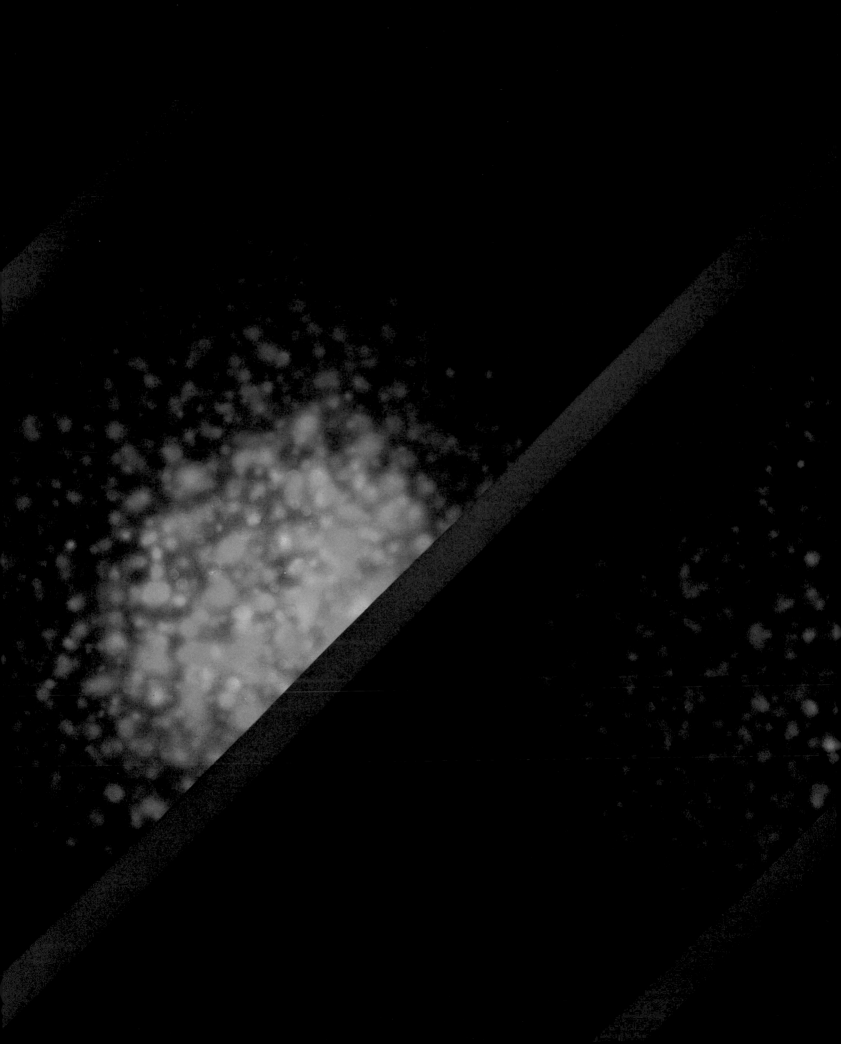

TURNING OUT THE LIGHTS

In a cosmos dominated by giant ellipticals, the supply of gas and dust for forming new stars will dwindle in all galaxies as each generation of stars returns progressively less material to the interstellar medium. Spent stars end their lives in one of three ways. Sun-size stars contract into white dwarfs, compressing a stellar mass into a body with roughly the diameter of Earth. White dwarfs will slowly radiate away heat until they cease to give off any light at all, becoming objects known as black dwarfs.

More massive stars are subject to even greater gravitational compression. Those with more than ten solar masses lose much of it to interstellar space when they go supernova, and then collapse into a body about ten miles across, so dense that subatomic particles coalesce into neutrons. Some of these neutron stars, known as pulsars, spend roughly 10 million years giving off pulses of radiation before eventually winding down. If the original star is larger still, it becomes a black hole with an event horizon two or three miles across, an object so massive that its gravitational pull allows nothing, not even light, to escape. About a hundred trillion (10^{14}) years from now, stellar evolution will have run its course, and galaxies will be little more than shadowy collections of dead stars held in the thrall of gravity.

At the Mercy of Passing Stars

Even after they die, stars will continue to orbit the galactic center much as they do today, and those that possess planetary systems will retain some part of their entourages for many millennia. Because the stars will have lost a significant amount of mass, however, their gravitational hold on their remaining planets will be considerably loosened. The planets will orbit farther from their home suns, making them vulnerable to the pull of other stars passing by. A single such encounter would typically do no more than slightly reshape a planet's orbit, but repeated jostlings could eventually kick a planet out of the system and ultimately strip a star of all of its satellites.

A star's velocity as it orbits the galaxy is another factor in the pace at which it loses its planets. A fast-moving star will encounter other stars more often than one traveling more slowly. However, since each encounter is fleeting, its influence is somewhat lessened, and more encounters would be required to knock a planet out of orbit.

Illustrated schematically at left is the effect that the passage of other stars would have on a planetary system in a dead galaxy a hundred thousand trillion (10^{17}) years from now. Two stars represent the hundred or so that would actually be needed to accelerate the planets to escape velocity—enough to break free of the gravitational bondage of their home sun.

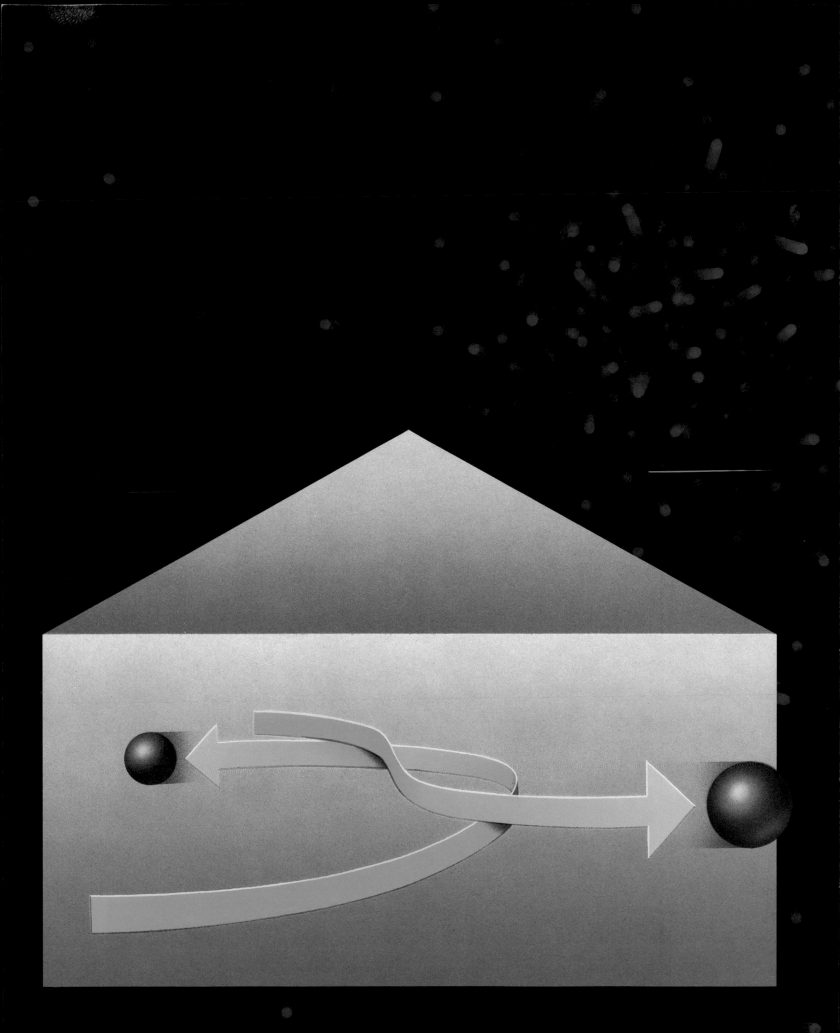

Galactic
Evaporation

Not only will close encounters between the dead stars in a galaxy strip planetary bodies from their orbits but, a million trillion (10^{18}) years hence, they will also strip the galaxy itself of stars, a process known as evaporation. Generally speaking, the farther a star orbits from the galactic center, the weaker the galaxy's gravitational hold; but the combined force of all other masses in the system is enough to prevent the star from flying off into intergalactic space. When two stars pass within about 10 million miles of each other, however, the encounter can alter their paths substantially, as illustrated at left.

The gravitational interaction between two stars moving in opposite directions transfers kinetic energy from one star to the other. The shift lowers the velocity of one star *(far left)*, thereby shrinking its orbit and causing it to move closer to the galactic center. The other star, for its part, receives an energy boost that is large enough to accelerate it to the velocity needed to escape from the galaxy.

More than 90 percent of the mass of a typical galaxy will evaporate in this way, departing for the vastness of intergalactic space. As the remaining stellar bodies fall into orbits that are more tightly bound to the galactic center, their mutual gravitation draws them into ever more frequent encounters with one another. Over time, neutron stars and black dwarfs will coalesce into black holes, which will then in turn grow into larger and larger black holes and, ultimately, into a black hole at the galaxy's center, if one did not already exist. That gravitational monster will then gather up any remaining matter into a single stupendous mass equaling 10^8 Suns.

The Slow Spiral to Immobility

Some of the stars cast out from their home galaxy carry their planets with them into the void, far enough away to avoid the irresistible suction of the galaxy's central black hole. But this reprieve is only temporary. The general theory of relativity predicts that planetary survivors will ultimately fall victim to a process that seems slated to destroy even the smallest relics of galactic structure.

The universe described by the general theory is likened to a four-dimensional fabric in which space and time are indivisible. This space-time fabric is sometimes represented as a two-dimensional elastic sheet. Level where there is no matter, the sheet is warped by any kind of mass, forming a depression called a gravity well. Very massive bodies such as stars make relatively deep gravity wells; planets, asteroids, and other small bodies dent the sheet only a little.

Because the path of a celestial object must follow the contours of space-time, a planet bound to a star orbits within the star's gravity well, as shown in the diagram at left. The tiny dimple that the planet itself makes in the space-time sheet moves with it, perturbing the gravitational field of the star-planet system. As the planet orbits, the perturbations, known as gravity waves, radiate outward at the speed of light, like ripples in a pond. Each wave carries away a minuscule amount of the planet's orbital energy, decreasing its velocity and causing the orbit to shrink. Moving in a slow downward spiral through the star's gravity well, the planet will eventually reach bottom and crash into the long-extinguished star. A trillion trillion (10^{24}) years from now, gravitational radiation will have brought all orbital motion in the universe to a halt.

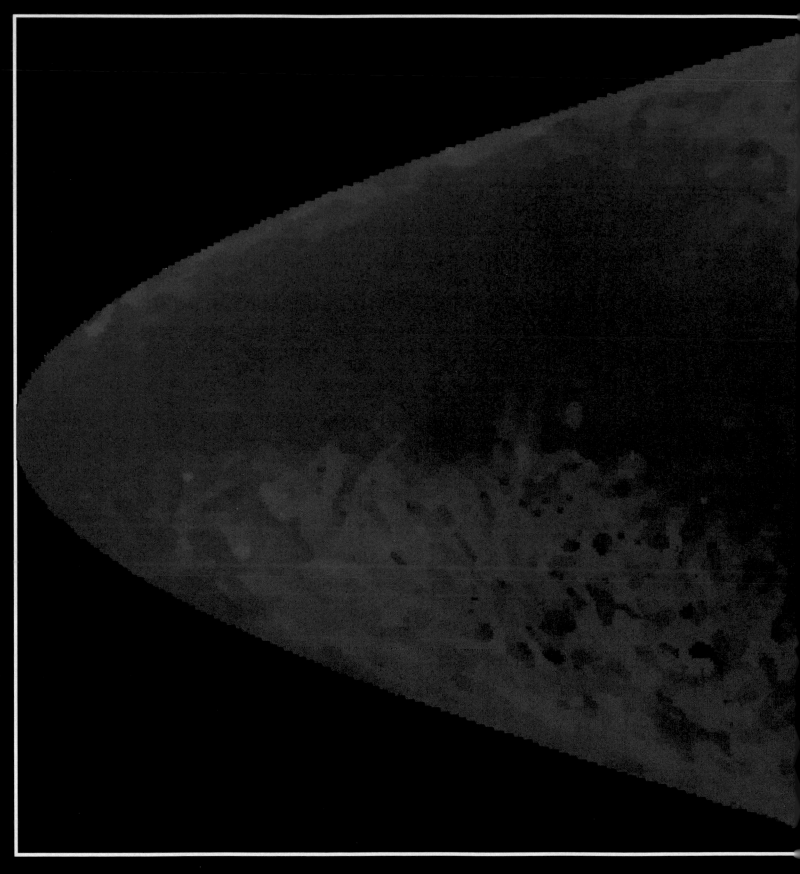

Cool radiation pervades all of space in this full-sky map of microwave emissions recorded by the Cosmic Background Explorer satellite early in 1990. The swath of purple indicates the radiation's remarkable evenness; pink and blue areas are distortions caused by the motion of the Milky Way against the cosmic background. At a uniform temperature of 2.7 degrees Kelvin, this background energy testifies to the expansion of the universe after its fiery birth in the Big Bang.

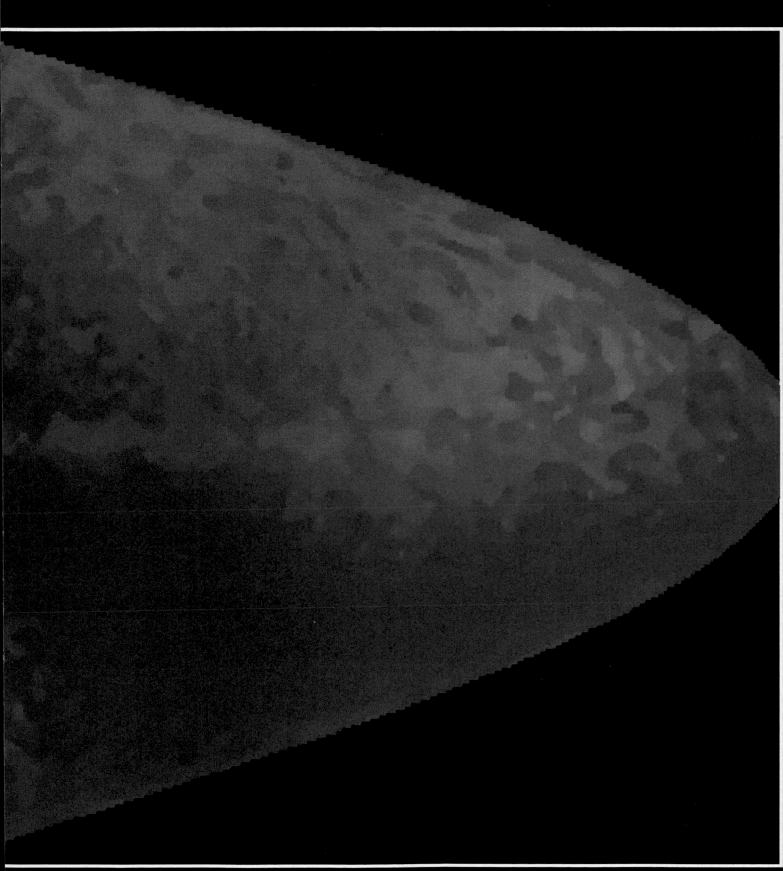

According to the laws of classical physics, the death of stars and collapse of galaxies into gigantic black holes would conclude the evolution of the universe. Never afterward could there be even a glimmer of light in the cosmos—only whispers of radiation growing ever feebler as the eons stretched into the infinite depths of time. Yet it seems inappropriate, somehow, that the universe, which began with the inconceivably energetic fireball of the Big Bang and which is constantly surprising astrophysicists with the violence of its processes, should meet such a dismal and anticlimactic end. As it happens, there are loopholes in the classical laws, discovered as physics experienced a revolution in the twentieth century.

Perhaps the most disconcerting finding is that matter and energy at the level of the atomic and subatomic world do not behave according to anything resembling common sense. In the realm of the very small, the cause-and-effect rules of normal existence give way to the peculiar precepts of quantum mechanics. As a result, dead stars need not necessarily stay dead, and black holes, once thought to be black forever, can—and, in fact, must—become white-hot, given enough time.

Whether or not sufficient time will be available depends upon effects operating on a scale as large as the quantum world is small—the scale of the cosmos itself. At present, the universe is expanding, as it has been for all of its history. It has grown from a seed far smaller than an atom to a dominion so great that light, moving at 186,000 miles per second, would take at least 40 billion years to cross from one side to the other. The expansion may continue forever; or it may slow down under the mutual gravity of all the contents of the universe, eventually halting altogether and giving way to a phase of contraction that will culminate in a fireball every bit as hot and dense as the original one.

The issue is whether the components of the universe have achieved "escape velocity" relative to each other. In rocketry, that term refers to the speed a missile must reach in order to escape Earth's gravitational pull by its momentum; without escape velocity, the rocket must inevitably crash back to Earth. In cosmology, escape velocity is the speed at which galaxies must fly apart in order to ultimately break their gravitational bonds. The problem is that while astrophysicists can calculate with high precision just how fast the galaxies are moving apart, they do not yet know the universe's total mass, and thus the strength of its gravitational field. And they do not yet have a reliable yardstick for cosmic distances—another important factor in the equation. These issues are some of the most hotly debated in modern astronomy—appropriately so, since the fate of the universe rides on them.

A NEW UNIVERSE

The realization that the universe is expanding came as something of a shock when astronomers first stumbled on the fact in the mid-1920s. Isaac Newton, the first great theorist of gravity, had assumed that the cosmos is static and unchanging. In particular, he believed the universe to be infinite in size and uniformly dense, which he thought implied that gravity should pull on all the stars with equal strength in all directions, thus canceling itself out and yielding a cosmos of satisfying permanence.

Three centuries later, the idea of an unchanging universe was still predominant. Even Albert Einstein, for all his radical thinking, was wedded to this view. He went so far as to deny the possibility of a nonstatic cosmos when the arguments in favor of it came from his own general theory of relativity, an immensely powerful and mathematically elegant description of the nature of space, time, and gravity. In working out the consequences of his equations in 1917, he discovered with alarm that expansion or contraction of the cosmos was inevitable. Yet a changing universe was so inconceivable to him—and so contradictory to all observational evidence—that Einstein made what he later referred to as the greatest blunder of his life. He marred the beauty of his theory by introducing an adjustment factor he called the cosmological constant. This extra number had the effect of a repulsive force that countered the overall gravitational field of the universe. It permitted a static solution to the equations.

Soon, however, others faced the facts that Einstein could not. In 1922, a Soviet mathematician named Alexander Friedmann, deliberately disregarding the evidence and assumptions of the day about the actual state of the universe, probed the relativity equations to see what possible solutions there might be. Specifically, Friedmann found that regardless of the value of the cosmological constant—even if it were set to zero—it was mathematically possible to model a universe that is expanding infinitely, as well as one that undergoes periodic episodes of expansion and contraction. A few years later, a Belgian priest and cosmologist named Georges Lemaître pursued similar ideas, quite unaware of Friedmann's work. It turned out that the smallest of perturbations could tip the scales, forcing the cosmos to grow or shrink. The situation was analogous to that of a pencil balanced perfectly on its point: In theory, the pencil could remain balanced indefinitely, but in practice, the slightest zephyr or vibration would knock it over.

So far, the debate had been conducted in the abstract realm of mathematics, but in the late 1920s, the American astronomer Edwin Hubble used observational evidence to shatter any lingering attachment to an unchanging universe. Hubble started out in life to be a lawyer, which was the same profession his father had followed. After graduating from the University of Chicago with a joint degree in mathematics and astronomy, he studied law at Oxford on a Rhodes scholarship, then opened a law office in Louisville, Kentucky. But he soon turned back to science, and in 1919 he went to work at Mount Wilson Observatory in California. His first major achievement there came in 1923, when he proved that the so-called spiral nebulae many astronomers thought to be located within the Milky Way were in fact full-fledged galaxies themselves, lying far out in the void. For this discovery alone, Hubble would have earned a place of high honor in astronomical history. But he was soon to accomplish much more.

Even before these fuzzy patches had been identified as galaxies, astronomers had begun analyzing the spectra of the nebulae—the rainbowlike array of frequencies that make up their electromagnetic radiation. The scientists hoped to find clues to the chemical elements the nebulae contained; experiments in earthbound laboratories had shown that particular frequencies in the spectrum of any light source are absorbed as the light encounters different types of atoms and molecules. But the lines in the spectra of the galaxies beyond the Milky Way, while clear enough as chemical signatures, were located in unexpected places along the frequency rainbow. In a few cases, the lines were all shifted toward the blue end of the spectrum, but generally the shift was redward. A well-known physical principle could account for such shifting. According to the law known as the Doppler effect, any wavelike phenomenon—whether a train whistle or radiation from a star—will shift to a lower frequency if the object transmitting it is moving away from the observer, and to a higher frequency if it is approaching. But why were most galaxies moving away from the Earth, as their redshifts attested? And why, as the degree of those redshifts indicated, did the velocities of recession increase with the apparent distance of the galaxies from Earth?

Although Hubble himself was reluctant to draw conclusions when he published his findings in 1929, the astronomical community soon acknowledged the obvious but entirely revolutionary explanation: The very fabric of space-time is expanding, in the process carrying all of the contents of the cosmos away from one another, like raisins in a fast-rising loaf of raisin bread. Distant galaxies are seen moving away faster than nearby galaxies because more space lies between them and Earth—more rising bread. The overall expansion is uniform, however.

The concept of a growing universe was accompanied by another extraordinary notion: The cosmic growth had to begin at some point back in time. Deep in the past, everything in the entire universe must have been contained within a tiny seed of incredibly concentrated energy and density. In 1931, Georges Lemaître envisioned this starting point of creation as a seething "primeval atom," which gave birth to the universe in an enormous explosion that would later be named the Big Bang.

More than a decade and a half later, in 1948, theorist George Gamow at

ALTERNATE FUTURES

Three possible fates await the cosmos, as illustrated in the diagram at right, which represents time horizontally and the size of the universe vertically. The deciding factor is total mass. Below a so-called critical density of the equivalent of three hydrogen atoms per cubic meter, the universe cannot generate enough gravity to hold itself together and expands forever *(top line)*. At precisely the critical density, it continues to grow in ever-smaller increments *(middle line)*, approaching zero velocity. In a universe with more than the critical mass, gravity eventually overcomes expansion and causes space to contract *(bottom line)*. Because its actual density is uncertain, the present expanding universe *(yellow mark)* could be following any one of these trails.

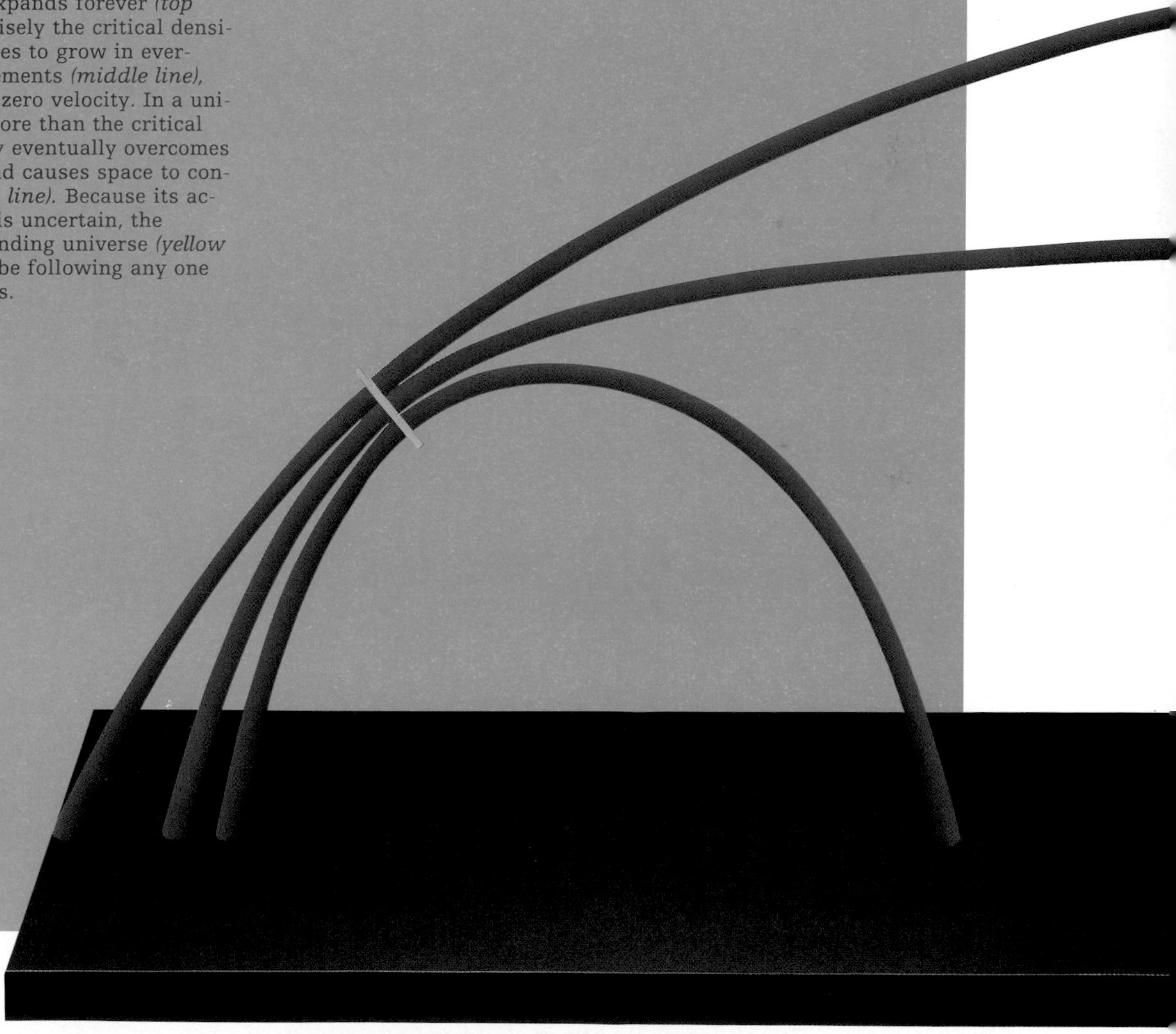

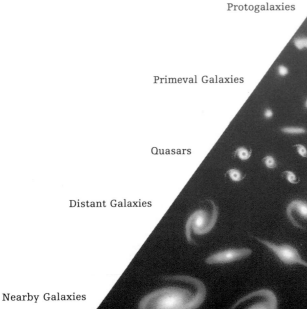

Big Bang

Ionized Hydrogen and Helium

Cosmic Microwave Background

Atomic Hydrogen
and Helium

Because light takes time to travel, far-distant
stars and galaxies appear now as they were
long ago, and the deeper scientists peer into
space, the farther back they see in time. As illustrated here, the view from Earth thus offers
a lesson in cosmic evolution. Beyond the stars
of the Milky Way *(bottom),* galactic history unfolds in reverse, all the way to the era of protogalaxy formation. The universe's initial stages
are even more remote; astronomers can see only
as far back as the era when radiation began to
propagate through space. The cosmic microwave background is the remnant of this early
light; previous events, including the Big Bang
itself, lie beyond detection.

Protogalaxies

Primeval Galaxies

Quasars

Distant Galaxies

Nearby Galaxies

Milky Way
Stars

George Washington University in Washington, D.C., and colleagues Ralph Alpher and Robert Herman at Johns Hopkins University in Baltimore published papers investigating how matter was formed in the Big Bang. But matter was only a small part of the story. The chief constituent of the newborn universe was energy—and the physicists suggested that the cosmos is still pervaded by its original electromagnetic endowment. Because the initial high frequency of the Big Bang energy has been stretched by the expansion of space-time, however, this radiance should now have a low-frequency, low-energy form. Expressing their energy estimate in terms of temperature, Alpher and Herman said that the fossil radiation should correspond to the microwaves emitted by a body at about five degrees above absolute zero, or five degrees Kelvin.

Remarkably, no one bothered to hunt for the remnant radiation, and the idea lay fallow for almost twenty years. Then, in 1965, Arno Penzias and Robert Wilson, two young scientists at Bell Telephone Laboratories in New Jersey, found it accidentally, in the course of a more mundane search for the source of some noise being picked up by a new microwave-receiving antenna. They tried everything they could think of to eliminate the problem, even cleaning out pigeon droppings that had accumulated in the antenna, but the noise—a reading that implied a body radiating at three degrees Kelvin—persisted. Meanwhile, just a few miles away at Princeton University, physicists Robert Dicke and James Peebles, unaware of the earlier prediction by Alpher and Herman, were on the track of fossil radiation from the birth of the universe, which Peebles had calculated should be at about ten degrees Kelvin.

Dicke and Peebles had already persuaded some Princeton colleagues to begin searching for the echo when they heard about the noise that was causing such frustrations for Penzias and Wilson. The microwaves picked up by the new antenna turned out to be coming from every direction in the sky with equal intensity—strong evidence of a Big Bang origin.

WILL THE EXPANSION CONTINUE?
With the discovery of the three-degree microwaves (also called the cosmic microwave background), the Big Bang seemed a scientific certainty. Astrophysicists began devoting whole careers to the exploration of its consequences, including the all-important question of how long the expansion of the universe will continue. A universe whose expansion goes on forever is called "open"; one whose expansion will slow, stop, and go into reverse is called "closed." There is also a third possibility: If the universe is somehow perfectly balanced between being open and closed, the expansion will continue unbounded into eternity, but at a rate that will get slower and slower, drawing ever more closely to zero. These ideas were already embodied in Alexander Friedmann's deliberations; he and his intellectual descendants found solutions for Einstein's equations that described each kind of universe.

Only one solution can be correct, though, and even today, nearly seventy years later, astronomers are just beginning to acquire the crucial data that may help them finally decide whether the universe is open, closed, or delicately balanced in between. The problem can be presented in two ways, essentially equivalent. One way is to ask whether the galaxies are moving fast enough to escape. The other is to ask whether there is enough matter in the cosmos, packed tightly enough, to keep them from doing so.

To resolve these issues, astronomers need to consider several parameters, all interconnected. The first is the relationship between the distances and recessional velocities of galaxies. Knowing a galaxy's redshift gives astronomers its actual speed away from Earth with great precision. It also supplies a measure of the relative distances of galaxies, for a simple reason: In a uniformly expanding universe, a galaxy with twice the recessional speed of another is twice as far away. The analogy of a rising loaf of raisin bread makes this clear. Consider that the entire loaf is growing at a constant rate—say, 10 percent a minute. The distance between any two raisins therefore also grows by 10 percent a minute. Next, imagine an observer on a "home raisin" looking out and seeing two other raisins, the first an inch away and the second two inches away. In a minute's time, each will have moved 10 percent farther away from the observer. But, in absolute terms, the nearer raisin has traveled a tenth of an inch in that minute, while the farther one has gone two-tenths of an inch, meaning that it has traveled twice as fast.

In practice, recessional speed can reveal the relative distances of galaxies, but it does not divulge actual distances. The cosmic relationship between actual distance and velocity is embodied in a number known as the Hubble constant, the rate at which the universe is expanding. This constant is usually expressed in the cumbersome-sounding speed-and-distance terms of kilometers per second per megaparsec. (A megaparsec equals approximately 3.26 million light-years.) If the Hubble constant were known precisely, it would be a simple matter to look at a galaxy's redshift, calculate its speed, and translate that into distance.

But astronomers do not know the Hubble constant with a high degree of accuracy; estimates place it anywhere between 50 and 100 kilometers per second per megaparsec. The reason for this lack of precision is that cosmic distances cannot be measured directly. Astronomers can gauge the distance to nearby stars quite accurately by means of parallax, the apparent shift in their position against the celestial background as Earth orbits the Sun. With these measurements in hand, they can estimate the distance to more remote stars in the galaxy—especially a type of star known as a Cepheid variable. Cepheids, it turns out, vary in brightness at highly regular intervals, and these intervals correspond to the stars' inherent average brightness. Thus, all an observer need do is watch the pattern of a Cepheid's variability to know how bright it actually is. During the light's travel to Earth, the actual brightness will be reduced according to the so-called inverse-square law of intensity: A star that is twice as far away as another star will look one-fourth as bright. Using that law, an observer can compare a Cepheid's intrinsic brightness to its apparent brightness and calculate its actual distance.

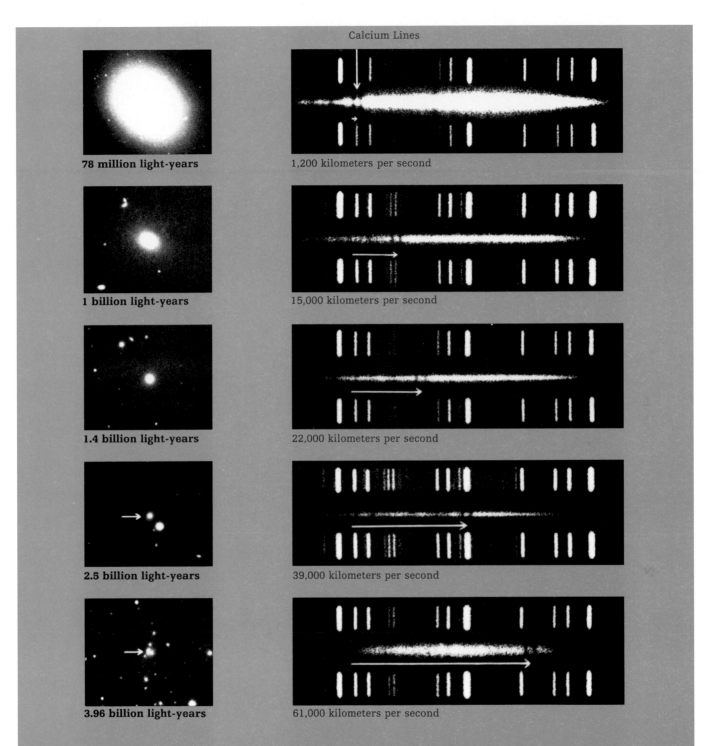

Calcium Lines

78 million light-years

1,200 kilometers per second

1 billion light-years

15,000 kilometers per second

1.4 billion light-years

22,000 kilometers per second

2.5 billion light-years

39,000 kilometers per second

3.96 billion light-years

61,000 kilometers per second

TELLTALE RECESSION

A sure sign of the universal expansion is that the more remote a galaxy is, the faster it is receding from Earth. The five elliptical systems at left, shown at the same magnification to reveal their increasing distance, illustrate the point. In each galaxy's spectrum (the hazy band between reference spectra above), a pair of dark lines denoting calcium absorption has shifted to the right *(horizontal arrow)*, a result of wavelengths being stretched as space expands. The amount of shift translates into a velocity directly proportional to distance.

It was by spotting a very faint Cepheid that Edwin Hubble confirmed the Andromeda nebula's location well outside of the Milky Way. The same clue has been used in estimating distances to a number of other galaxies, and these readings have provided the foundation for yet another distance-gauging trick: Examining many different clusters of galaxies, astronomers find the most luminous galaxy in each and assume that these are about equally bright in absolute terms; they then compare different degrees of observed brightness to arrive at the relative distances of the galaxies.

As the measurement process leads scientists ever farther into the universe, they can begin relating distance to redshift. (Closer in, the relationship is confused by the movements caused by local gravitational influences.) Each of the steps is reasonably accurate, but not perfectly so, and the small errors along the way add up to big errors in the end. Despite this handicap, astronomers are quite willing to choose their own preferred values for the Hubble constant, within the accepted range, and they can handily justify their choices as well. But the bottom line is that nobody really knows; the best astronomers can do is agree that the light from the most distant objects we see has been traveling for some 10 to 20 billion years, implying a universe that is at least that large.

COSMIC DENSITY

A central issue in the open-or-closed-cosmos problem is the density of matter and energy in the universe. Density determines the overall gravitational field of the cosmos and hence is intimately connected with escape velocity. Considering how fast galaxies are actually moving—the more distant ones are sailing away from Earth at an appreciable fraction of the speed of light—it might seem surprising that the average density needed to just close the universe (what astronomers call the critical density) is the equivalent of about three hydrogen atoms per cubic meter of space. Yet the cosmos contains so much matter in total that this suffices.

Exactly how much matter exists in the cosmos has yet to be determined—but cosmologists are working hard on the problem, which appears to be the most promising of all the approaches to deciding whether the universe is open or closed. The question is complicated by the fact that galaxies are evidently surrounded by invisible dark matter, stuff of unknown composition that astronomers have detected by its gravitational influence on individual galaxies' rotational speed and on their motion in relation to nearby star systems. In any case, the estimated total of dark matter and all visible matter falls short of the amount needed to close the universe. The total density detected in and around galaxies and clusters is only between 10 and 20 percent of the critical density. Still, there may be more dark matter lurking undetected in the vast regions between clusters of galaxies and scattered throughout the cosmos, perhaps in the form of unidentified subatomic particles.

While most astrophysicists, theorists and observers alike, are wrestling with the factors that might resolve the open-or-closed-universe issue, a few are focusing on the far future. What exactly will happen if the universe turns out to be closed, and on what time scale? What about an open universe—what will it look like in the depths of time?

Such luminaries as Einstein, Lemaître, and Gamow considered these questions in a general sense, but they lacked the information to go into them in any detail. In fact, the study of deep time did not really get under way until the 1970s, when the distant frontiers of time were reconnoitered by such scientists as Jamal Islam, then of Cambridge University in England, Freeman Dyson, and Steven Weinberg of Harvard University.

Dyson reasoned, for example, that a truly advanced technological civilization would find the ultimate way to tap energy from its parent sun: build a hollow sphere completely surrounding the star and live on the inner surface, where every bit of available solar energy could be intercepted and trapped. The trademark of such a "Dyson sphere" would be a faint glow of infrared energy as the sphere radiated away waste heat. That, argued Dyson, is what searchers for extraterrestrial intelligence should be looking for.

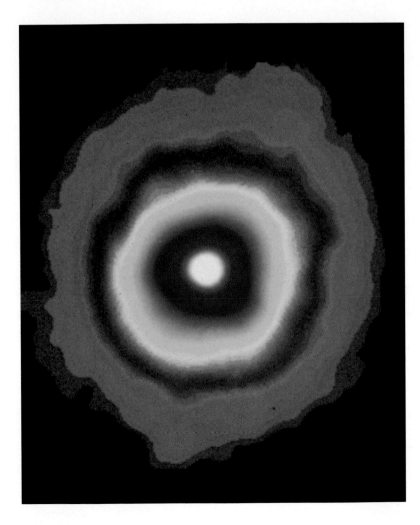

In this false-color x-ray image of the M87 galaxy, only the area shown in white would be visible at optical wavelengths. The other colors reveal the presence of hot gas that is gravitationally bound by vast amounts of undetected dark matter—as much as ten times the mass of the visible galaxy and twenty times that of the x-ray-emitting gas it is binding.

Steven Weinberg, whose scientific education began at the Bronx High School for Science in New York, has wrestled with some of the most mind-boggling aspects of cosmology. In his 1977 book *The First Three Minutes*, he describes in detail what conditions may have been like in the very early cosmos. Although his prime focus is on genesis, Weinberg has also examined the ultimate consequences of creation. Drawing on and extending Weinberg's work, Jamal Islam pursued a more thorough investigation of the implications of the universe's strange primeval physics.

FATE OF A CLOSED UNIVERSE

While the best current measurements seem to imply that the universe probably does not have enough matter to halt the expansion, a good scientist knows that probabilities are far from certainties. A paper published by Islam in the *Quarterly Journal of the Royal Astronomical Society* in 1977 described the less-likely scenario.

How long the cosmos would take to gradually grind to a halt and begin its collapse— slow at first, then faster and faster—depends on its endowment of matter. For the sake of argument, Islam said, imagine that there is just twice the amount needed to close the cosmos; that is, suppose that the actual density of matter is twice the critical density. If so, then the present expansion will last about 50 billion more years, more than twice the current age of the cosmos.

The galaxies will still be intact when expansion ceases, but their makeup will have changed considerably. Stars the size of the Sun will have ballooned to red giants, then settled into their old age as white dwarfs, while white dwarfs will cool to black. Stars with more mass will collapse into neutron stars and black holes. The supermassive black holes thought to exist at most galactic centers, patiently gobbling up whatever has come their way over the eons, will be even more massive. Only red dwarf stars, burning their nuclear fuel at a miserly rate, may survive largely unchanged. The cosmic microwave background will still be detectable, now cooled by the expansion of the

PIONEERING THEORISTS

Before the twentieth century, the standard assumption was that the dimensions of the universe were both measureless and unchanging. As far back as the late 1600s, however, Isaac Newton laid the groundwork for a scientific revolution that would ultimately dash all notions of a static cosmos. Often pushing at the limits of human imagination, the pioneering theorists shown here and on the following pages have unveiled an increasingly detailed portrait of the dynamic realities of an evolving universe.

1917 Unhappy that his theory of general relativity implied an expanding universe, Albert Einstein introduced an unknown constant into the equations to maintain cosmic stasis.

1687 Isaac Newton published the first theories on universal gravitation, setting the stage for later challenges to the concept of an unchanging cosmos.

1922 Russian mathematician Alexander Friedmann demonstrated the near impossibility of a static universe and calculated models that incorporated expansion.

universe to a chilly 1.5 degrees Kelvin. And the universe will measure some 120 billion light-years across.

The cosmos will be a rather dreary place, but civilizations presumably will have found ways to stay alive—burrowing deep into planets to conserve heat or building Dyson spheres to catch the feeble radiation from the few glowing embers of stars. In that long twilight, the exact moment when cosmic expansion stops and contraction starts may go unnoticed. True, the redshifts of galaxies will become blueshifts as recession turns to approach. But the light bearing the blueshifts will take time to arrive, millions of years even for the nearest galaxies, and far longer for distant ones.

However, as soon as any observers determine that the long-anticipated collapse has begun, they will know exactly how it will proceed, because gravity operates on the same timetable both forward and backward. A fly ball that reaches its apex in fifteen seconds will take just fifteen seconds to fall back to Earth; a universe that reaches its maximum size in 60 billion years will take the same amount of time to collapse, and it will go through the same stages as in the expansion, only in reverse.

There will be one difference, though: The stars, once burned out, will never spring back into life. Technically, stellar revival is not impossible—only so improbable that it amounts to the same thing. The reason is entropy. The

1927 Georges Lemaître, a Belgian priest and cosmologist, proposed an expanding universe that originated in an exploding fireball.

1929 Edwin Hubble related galaxy distances to recessional velocity—evidence of cosmic expansion.

1948 George Gamow, along with colleague Ralph Alpher, described a Big Bang origin of the cosmos and theorized that space should be awash in remnant energy at a temperature of five degrees Kelvin.

second law of thermodynamics is a simple proposition: In any closed system, the state of the system will evolve toward increasing disorder. In a closed room, for example, one might take all the air and compress it into a single corner, an orderly state, because the air is limited to a specific location. But if the air was left alone, the random motions of the individual molecules would spread the air throughout the room until it was evenly dispersed. That would be the state of maximum disorder, since any given molecule could end up anywhere. Nothing in the laws of physics prohibits these random motions from repositioning all the air back into the corner—and in fact, the air in any real room on Earth could do just that, leaving its occupants gasping for breath. But considering the trillions upon trillions of molecules involved, such a coincidence is so wildly improbable that it will never happen.

What applies to air molecules in a room also applies to energy in the universe. When a star dies, having dispersed its concentrated energy into space, it will not suddenly regather that energy and roar back to life. But that does not mean the collapsing universe will go on cooling. The microwave background radiation, which had been continuously weakening as the cosmos stretched, will start to warm up as the universe collapses. Fifty billion years after the collapse begins, the universe will have returned to its present size, and the microwave background radiation will have warmed up to its present three degrees Kelvin.

Ten billion years later, the cosmos will have diminished a hundredfold in size. Clusters of galaxies, containing tens and hundreds of individual islands

1965 Physicists Robert Dicke and James Peebles *(far left),* unaware of Gamow's work, attempted to detect ambient radiation left over from the infant universe.

1965 Astronomers Robert Wilson *(left)* and Arno Penzias stumbled upon this cosmic background radiation in the form of microwaves at about three degrees Kelvin.

of stars, will be so close together that they will merge into a single, universe-wide cluster. Meanwhile, the background radiation will have risen to 300 degrees Kelvin—about room temperature. The sky will be brighter with crowded stars and galaxies, but space itself will still be too cool even to radiate visible light.

Once 60 billion years of contraction have brought the universe to this point, the scenario will proceed much more rapidly. As the background radiation continues to heat, the space between galaxies and stars will no longer be dark, but will glow everywhere with light as bright as day. In a mere 10 million more years, the cosmos will have again shrunk by a factor of ten. It will then measure only 60 million light-years across—squeezing the mass of hundreds of billions of galaxies into an area that, at present, would contain a small handful. The background radiation will reach several thousand degrees, comparable to the surface of a star, and space will be so dense with matter and energy that the entire universe will seem to be enshrouded in a brilliant fog. In fact, at such a high temperature, the sky will be much too bright to look at; inherently cool bodies, such as planets, will have vaporized. In the crowded conditions of the cosmos, the giant black holes, which had grown ever more slowly as the expanding universe spread their fuel thin, will now voraciously gobble white dwarfs, neutron stars, and other black holes. The remaining white dwarfs and neutron stars will begin to vaporize like the planets.

Almost 300,000 years later—less than the lifetime of humanity—the cosmic background radiation will reach 10 million degrees Kelvin. Whatever is left

1979 Physicists Sheldon Glashow and Steven Weinberg *(bottom)* shared a Nobel prize with Abdus Salam for a theory unifying two fundamental forces in the first instants of the universe.

1979 Theoretical physicist Freeman Dyson published his hypothesis of the decay of ordinary matter in an ever-expanding universe.

1981 Cosmologist Alan Guth introduced the "inflationary scenario," in which many universes balloon out of a primordial cosmic medium.

of ordinary matter will boil off, the very atoms disintegrating into nuclei and electrons. The universe will be only 15,000 or so light-years across, about half the distance from Earth to the center of the current Milky Way.

In about a month, the collapsing universe will measure only ten light-years across, twice the present distance from Earth to Proxima Centauri, the nearest star in the Sun's neighborhood. The cosmic background temperature will have soared to 10 billion degrees. Even atomic nuclei cannot endure in these conditions; they will break into subatomic particles. Black holes will be gulping matter, energy, and each other at a prodigious rate—but not for much longer. The universe has only about one second left.

That is plenty of time, though, for the temperature to climb still higher. One-hundredth of a second before the end, the background radiation will reach a hundred billion degrees, and with about a tenth of a millisecond to go, it will hit one trillion degrees. Protons and neutrons will be torn apart into their constituent quarks, and the four forces of nature—gravity, electromagnetism, and the strong and weak nuclear forces—will start to combine. Physicists are convinced that the forces are not, at the deepest level, distinct at all; rather, they derive from a single underlying force that separated into four components as the universe cooled during expansion. Now, in the reverse situation, the first of the four to merge will be electromagnetism and the weak nuclear force. Then, as the temperature climbs through a thousand trillion trillion degrees, the strong nuclear force will combine with the other two. And finally, just .00000000000000000000000000000000000000001 second before the end of the universe, with the temperature rocketing toward infinity, gravity will join the other forces to produce a unified force whose parts are indistinguishable.

Beyond that point lies . . . the unknown. So far, the laws of physics have faithfully predicted what will happen to the universe. The timing of these events depends on what the density of matter in the universe is, of course, but the events themselves are, at least in principle, possible to determine. In the last instant of the drama, however, the universe is smaller than a subatomic particle, and at that level classical physics runs into trouble.

Up to this point, gravitational collapse has been satisfactorily described by the equations of Einstein's theory of general relativity; but on the minuscule scale the universe has now reached, quantum effects become important. Ordinarily, gravity has no relevance to the strange world of quantum mechanics, a world in which objects can be particles and waves at the same time, and where the exact position and momentum of a particle cannot both be known at once. Individual particles simply do not have enough mass for their gravitational interactions to matter.

But when the entire mass of the universe is squeezed to subatomic size, the situation is far different. Unfortunately, current theories for quantum effects and the workings of gravity are mathematically incompatible. Virtually all physicists believe it will eventually be possible to construct a so-called quantum theory of gravity; indeed, some of the brightest minds in the field are now trying to do just that. But until they succeed, this final instant of cosmic history must remain a mystery.

The mysterious nature of the end does not deter physicists from playing "what-if" games, naturally. As far as anyone knows, the gravitational field generated by an entire universe shrunk to subatomic dimensions is so great that the cosmos will collapse indefinitely, like the grandfather of black holes, approaching infinitesimal size and infinite density. On the other hand, instead of collapsing, the universe may suddenly begin to grow again—a notion prefigured by Alexander Friedmann in 1922. If that were to happen, there would not be a Big Crunch after all. The universe might really be a cyclic phenomenon in which the final moment of contraction obliterates all evidence of what came before, and a new Big Bang starts the whole thing over again. But perhaps not quite all the evidence of the previous universe is wiped out. In each cycle, stars produce electromagnetic radiation of various frequencies. As contraction progresses, the radiation would be increasingly compressed to higher frequencies. If this higher-energy radiation is conserved through the bounce, the subsequent cycle will have a longer lifetime—about twice as long, according to Princeton's Dicke and Peebles. The principle also works in reverse: Cycles of expansion and contraction would have been shorter in the past. In fact, just 100 cycles ago, there would have been barely enough time for the universe to form stars, let alone planets and galaxies.

Even more astonishing than vistas of a bouncing cosmos is the possibility that there might well be more than one universe, existing not in succession but simultaneously, a notion advanced by the Soviet physicist Andrei Linde in 1983. Linde's speculation depends in part on another theory, called inflationary cosmology, that describes a mechanism whereby the known universe of stars and galaxies grew dramatically from the Big Bang. Alan Guth, a physicist at the Massachusetts Institute of Technology, has calculated that under the enormous density that existed in the wake of the Big Bang, a powerful negative gravitational force could have arisen, causing the universe to balloon outward with tremendous rapidity. Furthermore, according to the laws of quantum physics, it is possible—even certain—that on immeasurably short time scales, energy can arise from nowhere and then disappear. The shorter the time scale, the greater the energy allowed. Putting these two concepts together, some physicists have argued that the Big Bang was a quantum fluctuation that got out of hand when the inflationary mechanism proposed by Guth kicked in with its negative gravitational force. But if that is true, asked Linde in his paper, why should there have only been a single Bang? The universe we inhabit, he suggested, might really be just a small part of a larger cosmos, in which mini-universes pop up all the time; the laws of physics might even be different in each. Moreover, each new universe would have the capability of spawning its own offspring, which would immediately disappear from the parent space-time continuum by just the same mechanism. Our own cosmos might someday end, but a web of other universes would stretch infinitely forward, backward, and even sideways in time.

AN OPEN UNIVERSE—WITH PROTON DECAY

Until someone can show convincingly that the universe does have a critical density of matter, physicists and astronomers must respect the existing evidence and assume that the expansion will go on forever. This merely exchanges one set of mysteries for another, however, because evolution in an ever-expanding direction also carries some large unknowns. Just how infinite expansion will unfold depends on whether or not certain hypotheses currently being worked out by theoretical physicists hold true. Known informally as GUTs—for Grand Unified Theories—they attempt to explain the relationships among three of the four forces of nature. (The fourth force, gravity, is on the list for inclusion as well, but physicists agree that incorporating it into a final, superunified theory will be extremely difficult.) The first GUT was devised back in 1974 by Howard Georgi and Sheldon Glashow of Harvard University. (Glashow eventually shared a Nobel prize with Steven Weinberg and Abdus Salam of Imperial College, London, for earlier work finding a quantum link between electromagnetism and the weak nuclear force.) That first rough effort at a unifying theory version carried with it a profound consequence: Given enough time, protons should decay, dissipating into a variety of subatomic particles *(pages 111-113).*

Physicists had long wondered whether the proton, a fundamental building block of atomic nuclei, would keep its form forever. They knew that free neutrons, the electrically uncharged cousins of protons, decay quickly—in a matter of minutes, in fact—into protons, electrons, and antineutrinos. But no one had ever detected a proton decaying, and no theory demonstrated that they should until Georgi and Glashow's first GUT.

Soon after the two scientists published their work, Steven Weinberg and physicist Helen Quinn calculated how long the average proton should live: 10^{31} years. This might seem a difficult prediction to test, given the time involved. But that proposed span of 10^{31} years is the average lifetime of a single proton. Virtually all protons will live either longer or shorter lives, according to the theory. Straightforward calculations showed that if scientists were to assemble 10^{31} protons—the number in a few hundred tons of water—they would see, on average, one decay per year.

That is exactly the principle behind proton-decay experiments now under way in a salt mine under Lake Erie, in a gold mine in India, and in a tunnel under Mont Blanc in the Alps. (These experiments are all being conducted underground so that cosmic rays—stray atomic fragments that bombard Earth from space—will not confuse the electronic detectors looking for the charged particles that testify to proton decay.) The verdict so far is that if the proton really does decay, its lifetime is at least 10^{32} years—a much longer span of years than the original Grand Unified Theory prediction but compatible with subsequent ones.

If protons do in fact decay, their release of not only particles but also energy means that the black dwarfs, dead planets, and neutron stars that

Quark **Antiquark**

Electron **Positron**

Photon

Leptoquark

Pi Zero

A subatomic roster. According to quantum theory, positively charged protons—along with neutrons, their neutral companions in the atomic nucleus—are made up of fundamental particles known as quarks. Negatively charged electrons are members of another family of particles known as leptons, and both leptons and quarks may have been spawned by leptoquarks. Quarks and electrons have corresponding antiparticles—antiquarks and positrons—that are equal in mass but opposite in other characteristics, such as charge. When matter and antimatter particles collide, they convert mass into photons of energy. Quarks and leptons can exchange identities, resulting in the massive but fleeting pi zero particle.

An Unstable Future

One of the basic tenets of quantum mechanical theories for the origin of the universe is that matter and antimatter were created in equal proportions during the earliest moments after the Big Bang *(inset, below)*. Given that assumption, however, the universe should not exist, because matter and antimatter would have annihilated each other, leaving only the radiation generated by that mutual destruction.

Physicists thus have had to posit a minute excess of matter—one extra particle per billion—to explain how the observable cosmos came to be. In the intricate and bizarre quantum world, this deceptively small adjustment has one consequence with enormous implications for the long-term future of the universe: Protons, subatomic particles once held to be stable, must decay *(page 113)*. And if that is in fact the case, all organized structures in the universe, from atoms to galaxies, will ultimately dissolve.

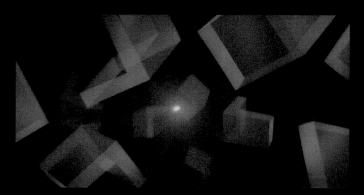

Leptoquarks that formed instants after the Big Bang are thought to have eventually given rise to both matter and antimatter—in the form of quarks, antiquarks, electrons, and positrons—which should have annihilated one another. To justify the existence of a matter universe, quantum theory allows for the generation of a one-particle-in-a-billion excess of matter.

THE GEOMETRY OF PROTON DEATH

Of several mechanisms physicists have devised for the decay of protons, the most-studied scenario is known as the "$e^+\pi^0$" process, so called for two products of the decay, a positron (e^+) and a pi zero particle (π^0). Experiments performed in high-energy particle accelerators have shown that the particles produced during the protons' disintegration would trace a distinctive pattern, flying away from each other at a precise angle, as illustrated here.

The $e^+\pi^0$ process begins with a proton *(top left)*, composed of three quarks of two different types: two so-called up quarks and one down quark. The two up quarks spontaneously merge to form a leptoquark. This very massive and short-lived intermediary then decomposes into two antimatter particles, a positron and an anti-down quark. As the positron flies outward in one direction, the anti-down quark combines with the leftover down quark from the original proton to form, for the briefest of instants, a pi zero particle. This particle, flung in the opposite direction from the positron, quickly disintegrates, leaving behind a pair of highly energetic gamma ray photons. The gamma rays, in their turn, may form electron-positron pairs whose members could then annihilate one another to produce photons.

have managed to escape being eaten by black holes will not be quite as cold and dead as one might imagine, a notion first proposed independently and almost simultaneously in the early 1980s by three groups of physicists: Don Page and Randall McKee, then of Pennsylvania State University (Page is now at the University of Alberta in Edmonton); Gerald Feinberg of Columbia University; and Duane Dicus and John Letaw of the University of Texas at Austin with Doris Teplitz and Vigdor Teplitz of the University of Maryland. According to their theories, each decaying proton should release a tiny bit of heat. They added to this the energy released by neutrons bound in a star (ordinarily stable), which decay by a similar process. Denser objects, containing more protons and neutrons in a smaller space, should by this process radiate more heat for a given surface area. They concluded that, assuming that the proton decays in 10^{32} years, the least dense of the dead stars in the universe, black dwarf stars, will hover at around one degree Kelvin, while the more compact neutron stars will be a relatively toasty thirty degrees Kelvin.

Page and Dicus, together with their colleagues, pushed their ideas even farther, delving into a question until then rarely addressed: What would happen next? They theorized that all ordinary matter in the cosmos—burned-out white and red dwarf stars, the free-roaming husks of planets—will be reduced to just electrons and their antimatter twins, positrons, wandering through a universe of giant black holes. Keeping them company will be neutrinos left over from the Big Bang and the star-burning era. Created in nuclear reactions, these ethereal subatomic particles have small or perhaps even no mass and are so indifferent to other matter that they can pass through light-years of lead without stopping. There will also be photons still carrying the remnants of the original microwave background radiation, now cooled to at least .000000000000001 degree above absolute zero.

The universe, which at this point will have been expanding for 10 billion trillion (10^{25}) times as long as the expansion has gone on so far, will be 10^{15} times its present size, and the average distance between individual particles will be greater than the size of the Solar System. Because electrons and positrons are oppositely charged, they will attract each other, albeit weakly, even over such great distances.

Back in 1978, physicists John Barrow and Frank Tipler of the University of California at Berkeley had suggested that in a universe balanced between open and closed, electron-positron pairs could latch onto each other and form "atoms" of what physicists call positronium, consisting of a single electron and a single positron orbiting each other. In 1980, Page and McKee refined Barrow and Tipler's theory and worked out the mechanism for the formation of positronium, showing that it would not take place until the cosmos was some 10^{75} or so years old. By that time, the expansion would have carried the orbiting electrons and positrons so far apart that each atom of positronium would be much larger than the entire visible universe is today.

And then the positronium atoms, too, would decay. It has been known since the middle of the nineteenth century that an orbiting electron continuously emits photons and loses energy; so does a positron. Gradually and imperceptibly, the particles will spiral in toward one another as they lose energy; finally they will collide in a microinferno that destroys both and emits a small burst of gamma radiation. By then, the cosmos will be at least 10^{124} years old: ten million billion billion billion billion billion billion billion billion billion billion billion billion years.

Even most black holes will have long since disappeared. On the face of it, this might seem absurd: A black hole contains matter compressed to virtually infinite density and infinitesimal size. What could possibly destroy such a thing? Even another black hole's swallowing it would simply lead to a single, more massive hole where once there were two.

The answer is found in quantum mechanics. In 1974, the great Cambridge University theoretician Stephen Hawking applied quantum precepts to black holes and came up with the paradoxical conclusion that black holes are not black. In fact, they glow with radiation. Hawking made use of the same quantum-mechanical phenomenon that allows universes to spring up from nowhere. On very short time scales, pairs of particles called virtual particles, one made of matter and the other of antimatter, can appear spontaneously anywhere in space, then disappear. But what would happen, he asked, if such a pair appeared right at the edge of a black hole? In some cases, one of the pair would be sucked down the black hole and the other would escape. That poses a problem: A new particle has suddenly appeared from nowhere. Hawking realized that, for the books to balance, energy must also disappear from somewhere—in this case, from the black hole, which loses a tiny bit of mass. The escaping particle appears to an outside observer to be energy radiating from the black hole. Eventually, this effect forces the black hole to disappear entirely. According to Hawking's figures, smaller black holes radiate energy and lose mass more quickly than bigger ones. The process, therefore, is an accelerating one: A black hole of any given size will get hotter and hotter and smaller and smaller until, at the end, it explodes in a brilliant burst of energy. Commented Freeman Dyson, "The cold expanding universe will be illuminated by occasional fireworks for a very long time."

The fireworks should begin when the cosmos is about 10^{67} years old, the time it would take a black hole with the mass of a star to radiate away all its energy. For supermassive black holes fashioned from a galaxy's worth of matter, the process will take a little longer—until the year 10^{90}, perhaps. And the very biggest black holes, fashioned from the matter of hundreds or thousands of galaxies, may not lose their mass until the universe is 10^{100} years old. Eventually, though, given an open universe, every last black hole will evaporate, leaving behind only an increasingly dispersed field of photons and other stable particles such as electrons, positrons, and neutrinos.

AN OPEN UNIVERSE—WITHOUT PROTON DECAY
But what if the GUTs are wrong, at least on the question of proton decay? The physics of positronium and of black holes will remain unaffected, as will their

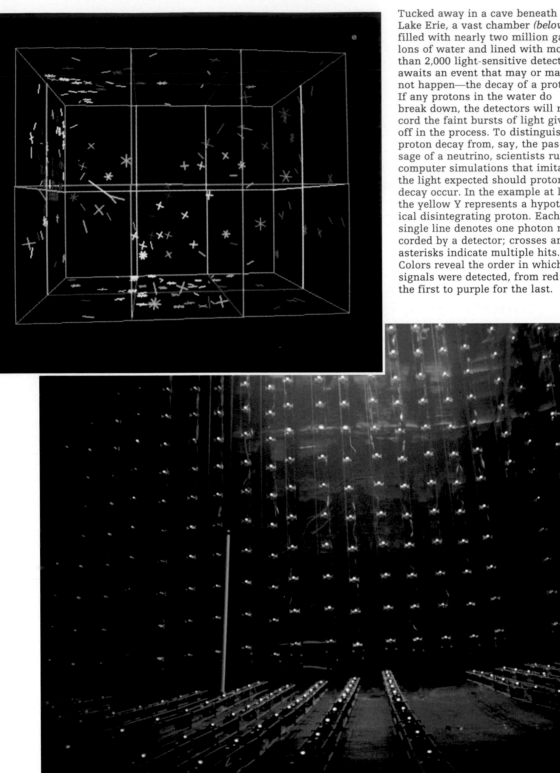

Tucked away in a cave beneath Lake Erie, a vast chamber *(below)* filled with nearly two million gallons of water and lined with more than 2,000 light-sensitive detectors awaits an event that may or may not happen—the decay of a proton. If any protons in the water do break down, the detectors will record the faint bursts of light given off in the process. To distinguish proton decay from, say, the passage of a neutrino, scientists run computer simulations that imitate the light expected should proton decay occur. In the example at left, the yellow Y represents a hypothetical disintegrating proton. Each single line denotes one photon recorded by a detector; crosses and asterisks indicate multiple hits. Colors reveal the order in which signals were detected, from red for the first to purple for the last.

projected lifetimes. However, ordinary matter, in the form of dead planets and burned-out or collapsed stars, will persist far longer.

Upon peering into these ultimate depths of time, Freeman Dyson has reasoned that the matter will not last forever. Quantum mechanics dictates that atoms in such supposedly stable configurations as solid rock will, if allowed enough time, spontaneously jump into states that are still more stable. Even at a temperature of nearly absolute zero, for example, solid matter will, over time, flow like a liquid. Under the force of gravity, this effect will force every chunk of solid matter into a perfect sphere within 10^{65} years.

Over much longer timespans, matter will change even more radically. In the short term, elements like uranium and plutonium are unstable, decaying by fission into lighter elements. But in the depths of time, quantum mechanical rearrangement will cause other elements, not normally considered radioactive, to do the same. Since the most stable element is iron, all elements heavier than iron will eventually decay into it by slow nuclear fission, while everything lighter will eventually fuse into iron. These processes will be complete, Dyson found, in $10^{1,500}$ years.

But quantum mechanics is not finished yet. Although iron is the most stable form of ordinary matter, there are extraordinary forms of matter as well. These already exist in the universe—though not on Earth—in the shape of neutron stars, composed of atomic nuclei packed so tightly together that a teaspoonful on Earth would weigh about a billion tons. The only way to create this nuclear matter in the present universe is through the intense gravity of a collapsing stellar giant. Over the eons of deep time, though, the packing will happen through quantum movements. As with any change from one form of matter to a more stable form, enormous bursts of energy will be released, adding to the feeble radiation bathing the cosmos. By the year $10^{10^{76}}$, virtually no ordinary matter would be left, just nuclear matter.

But Dyson suggests that this state of affairs will never be reached. Black holes are even more stable than neutron stars, and Dyson has calculated the time scale for conversion of all matter in the universe, including existing neutron stars, into black holes. He found that if only a small fraction of a dead planet, for example, rearranged itself into the density of a black hole, that fraction would immediately begin swallowing the rest, turning the entire object into a black hole in very short order. Thus, even though it is more difficult for ordinary matter to achieve black hole density than neutron star density, the black hole scenario would proceed much faster; indeed, black holes could spontaneously form in as little as 10^{45} years, and by $10^{10^{26}}$ years, only black holes would be left. Then, as was the case with the earlier generation of holes, these would finally evaporate, leaving only photons and other stable particles in a vast cosmos. Spontaneous quantum fluctuations will continue, of course, and so the universe will go on percolating below the level of observability, perhaps occasionally giving rise to a new universe that forms its own space-time and evolves along a separate route. Otherwise, there will be nothing but darkness.

THE PROSPECTS FOR LIFE

Considering that human beings live on a planet where life has been around for scarcely a few billion years or so, and in a universe where even the most ancient extraterrestrial civilization can be no more than 20 billion years old, it might seem presumptuous to talk about intelligence surviving into deep time. Yet Dyson, at least, has been willing to speculate about just that. The question depends, he argues, on whether or not consciousness must be rooted in the sort of matter that makes up brains and computers, or whether it can arise from particles other than those constituting atoms and molecules.

He would argue that such an evolution is possible as long as the other constituents can be organized into structures analogous to those that carry consciousness. Perhaps the photons that survive the death of black holes could somehow fashion a network that could carry out the functions of neurons in the brain or silicon switches in a computer. Operating at an energy level near absolute zero, and with its components separated by billions of light-years, this universe-spanning mind would of necessity take eons to complete a single thought. But, of course, in the infinite depths of time, there will hardly be any reason for hurry.

It may seem idle to speculate about the fate of the universe so far into the future, and perhaps it is. But such is the insatiable curiosity of the human species that if a question can even be posed, it will inevitably be explored, with whatever intellectual tools are at hand. As Steven Weinberg put it in *The First Three Minutes:* "The effort to understand the universe is one of the very few things that lifts human life above the level of farce, and gives it some of the grace of tragedy."

INTIMATIONS OF ETERNITY

The rate at which distant galaxies are everywhere receding is generally taken as strong evidence that the universe is expanding—and has been since its presumed birth in the violent explosion called the Big Bang. Whether that expansion will succumb in time to the forces of gravitational attraction is a huge question whose answer rides on a very small number, the so-called critical density of matter and energy in the universe: the equivalent of three hydrogen atoms per cubic meter.

If the average density of matter and energy is less than the critical value, then the universe is deemed open. With too little substance to check cosmic growth, it will continue to hurtle outward indefinitely, becoming colder, darker, and emptier through the eons. Alternatively, the density may exactly match the critical value, in which case the universe is flat; gravity will slow expansion but never quite stop it. Only if the density exceeds the critical value will gravity triumph. Held in its relentless grasp, the universe will eventually cease its outward motion and contract, heading for a catastrophic implosion known as the Big Crunch.

The scenarios offered on the following pages explore various evolutionary possibilities for the open, flat, and closed universe. Over stretches of time so endless that the concept loses all meaning, bizarre subatomic effects come into play. Under their sway, the universe may diffuse into a tenuous blanket of matter and antimatter debris; or its stars may turn to iron and then to black holes; or, in the end, the cosmos may devour itself.

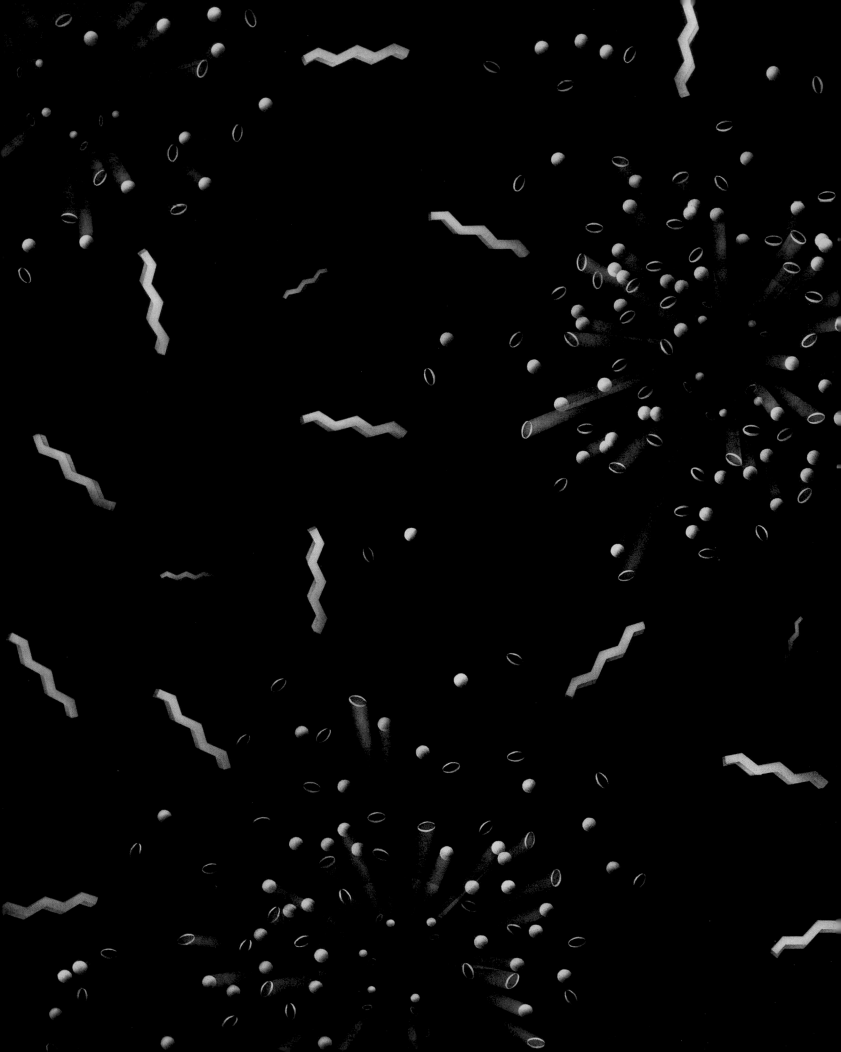

THE PROTON'S LEGACY

If the universe is open, a trillion trillion (10^{24}) years of expansion will find it strewn with the flotsam of galaxies long since dispersed: extinct stars, planets cast adrift, black holes. But even in this dark and frigid infinity, feeble waves of infrared heat may radiate from the surface of stellar remnants.

Such heat, the by-product of the so far speculative subatomic process known as proton decay *(pages 111-113)*, may maintain temperatures of thirty degrees Kelvin in massive stars and one degree Kelvin in lesser masses—scarcely measurable by present-day stellar standards, but fiery in the utter cold of that future cosmos. If protons do decay, so will neutrons, their companions in the atomic nucleus. The disintegration of protons and neutrons will liberate a host of particles: electrons and their antimatter twins, positrons, along with neutrinos and antineutrinos. Only the neutrinos and antineutrinos *(green balls and rings, left)* will escape from the interiors of stars into interstellar space. The more interactive electrons and positrons will annihilate each other during chance encounters, releasing high-energy photons that will warm the dead stars enough to emit infrared radiation *(white)*. As the stars dwindle in mass, more and more of the particles produced by proton decay will escape, until the stars have evaporated—a period lasting roughly 10^{32} years.

Proton decay will also transform the last vestiges of interstellar gas, converting it to an exceedingly rarefied plasma of electrons, positrons, and neutrinos. By the end of the era of proton decay, some 10^{33} years hence, an average of nine billion miles will separate these bits of matter. They in turn will share the ever-expanding reaches of space only with black holes and the incidental photon left over from the Big Bang, star shine, or proton decay.

Positronium's Strange Reign

In an open universe subject to proton decay, the particles released by the proton's disintegration will be so rapidly carried apart by the cosmic expansion that they will be unable to interact. If the universe is flat, however, electrons and positrons will remain close long enough to pair off, orbiting their common center of mass and forming colossal atoms that physicists have named positronium *(left)*.

Some 10^{75} years from now, after the last protons have decayed, the interlocking orbits of one bound pair will circumscribe a region of space roughly 10^{21} light-years in extent—an area far larger than today's observable universe. However, the cosmos will itself be so much vaster by then that a positronium atom will be, in effect, roughly a thousand times smaller than the period at the end of this sentence. Over the eons, orbital motion will rob the constituent electrons and positrons of energy, causing them to spiral inward. Eventually, they will collide and annihilate each other in a flash of high-energy photons *(white)*. According to one scientific theory, by AD 10^{124}, most of the positronium atoms will have self-destructed. But as the flat universe continues to expand at an increasingly slower pace, the dwindling number of positronium atoms—formed later and so tracing ever-larger orbits—will take longer and longer to decay. Positronium will reign supreme to the end of time.

Quantum Magic

If it is the case that protons do not decay, the evolutionary path of matter in the open universe will be profoundly different. Instead of disintegrating into a diaphanous particle veil, the stuff of the universe will undergo several alchemic transformations: first to liquid, then to iron, and finally—for larger masses, at least—to black hole oblivion.

The mechanism responsible for these strange metamorphoses is the so-called quantum tunneling effect. Under its influence, ordinary matter devolves to the lowest energy state through a subatomic particle exchange that permits electrons and other quanta to slip Houdini-like through classical energy barriers, rearranging chemical and physical structures in the process. Through such particle swapping, all aggregated matter will behave like a liquid by AD 10^{65}. Consequently, gravitational action will then cause even the most irregularly shaped rock and metal fragments to meld into spheres.

Some $10^{1,435}$ times as many years will pass before quantum tunneling performs its next magic trick: the conversion of all clumped matter to iron. Anything larger than about one ten-thousandth of an inch across—planets, moons, asteroids, and stellar ruins such as black dwarfs—will transmogrify into purest iron, the element possessing the most stable nucleus.

Then, in the deepest reaches of cosmic time—$10^{10^{76}}$ years from now—even the iron stars may crumble. The same subatomic processes that led to matter's chemical conversion to iron should cause stellar objects to seek still lower energy states: Iron black dwarfs *(left, purple)* will degrade into compact neutron stars *(blue)*, which in turn will collapse to form black holes *(purple circle)*. Energy liberated during the change from dwarf to neutron star will produce brilliant fireworks, bursts of x-rays and neutrinos reminiscent of supernovae. Such light shows will punctuate the darkness for untold eons, until at last a near-absolute blackness will prevail. Black holes, accompanied only by wandering bits of iron, interstellar particles, and stray photons, will stand sentinel over the declining universe, until, finally, they too will pass away.

Ephemeral Monsters

Though scientists differ on whether the fate of the open universe will be governed by proton decay or quantum tunneling, they all agree that the outcome will be much the same: Those final cosmic holdouts, black holes, will themselves decay.

The notion contradicts classic relativity theory, according to which such an eventuality is impossible: Every black hole is bound by a horizon whose intense gravity would prevent any matter or energy inside it from escaping. Were this true, black holes and their cache of collapsed galaxies and stars would endure for eternity. But quantum theory dictates that black holes are mortal after all: Ever so slowly—over as many as 10^{100} years—they will leak their energy to the surrounding void and evaporate.

The agents of their destruction are known as virtual particles—particle-antiparticle pairs that have the potential to materialize as ordinary particles so fleetingly as to escape detection. To do this, they must borrow energy from the cosmos, which they then "return" when annihilating one another. However, a pair of virtual particles—including photons, which act as their own antiparticles—could flash into existence near a black hole, and then become separated when one disappeared into the black hole abyss. The survivor, lacking its partner, cannot be annihilated; needing a source of mass to become "real," it takes a small amount of energy from the black hole's gravitational field and moves off into space. In this way, little by little, the black hole is robbed of energy and, consequently, of mass.

As the escaping particles and low-energy photons radiate into space *(far left, top and bottom)*, the black hole shrinks and grows hotter, until at last, unable to maintain its gravitational stranglehold, it explodes in a frenzy of high-energy gamma rays and particles *(left)*. Thus will it go until all the black holes have evaporated. Only wandering electrons, positrons, photons, and neutrinos will remain.

COSMIC REVERSAL

The odds no more favor an open or flat universe than they do one that is closed—one that, sooner or later, will cease to expand and begin to deflate like a balloon leaking air. If the universe is closed, the exact moment of reversal will depend upon the average density of matter: The nearer it is to the critical value of approximately three hydrogen atoms per cubic meter of space, the longer the period of expansion will last. Scientists believe that if the universe is closed, the actual and critical mass densities are unlikely to be so nearly congruent; more probable, they say, is a higher-mass, shorter-lived universe whose maximum expansion will last long enough to allow the disintegration of galaxies and the death of stars but not the demise of supermassive black holes.

No one can predict with certainty the final age of the closed universe, but certain clues will signal the beginning of the end. If the universe is closed, the rate of cosmic expansion will eventually slow down; the recessional velocities and corresponding redshifts of distant objects will begin to diminish until expansion stops altogether and contraction begins. As the collapse progresses, the long wavelengths of the cosmic background radiation—the greatly red-shifted remnants of energy unleashed by the Big Bang—will begin to shorten as they are compressed toward the blue, or higher-energy, end of the electromagnetic spectrum. The blueshifting will intensify and the radiation will become increasingly energetic *(right)* until, one year before the final crunch, the ambient temperature of space—once barely a millionth of a degree above absolute zero—will be as hot as the surface of today's Sun. Bathed by the 100-million-degree plasma, burned-out stellar relics such as black dwarfs and neutron stars will vaporize or explode, spewing their superheated particles into the shrinking cosmic furnace. Black holes *(dark spots)* feeding on the radiant debris and on some whole stars before they vaporize will expand exponentially even as the universe is squeezed smaller by the inexorable vise of gravitational collapse.

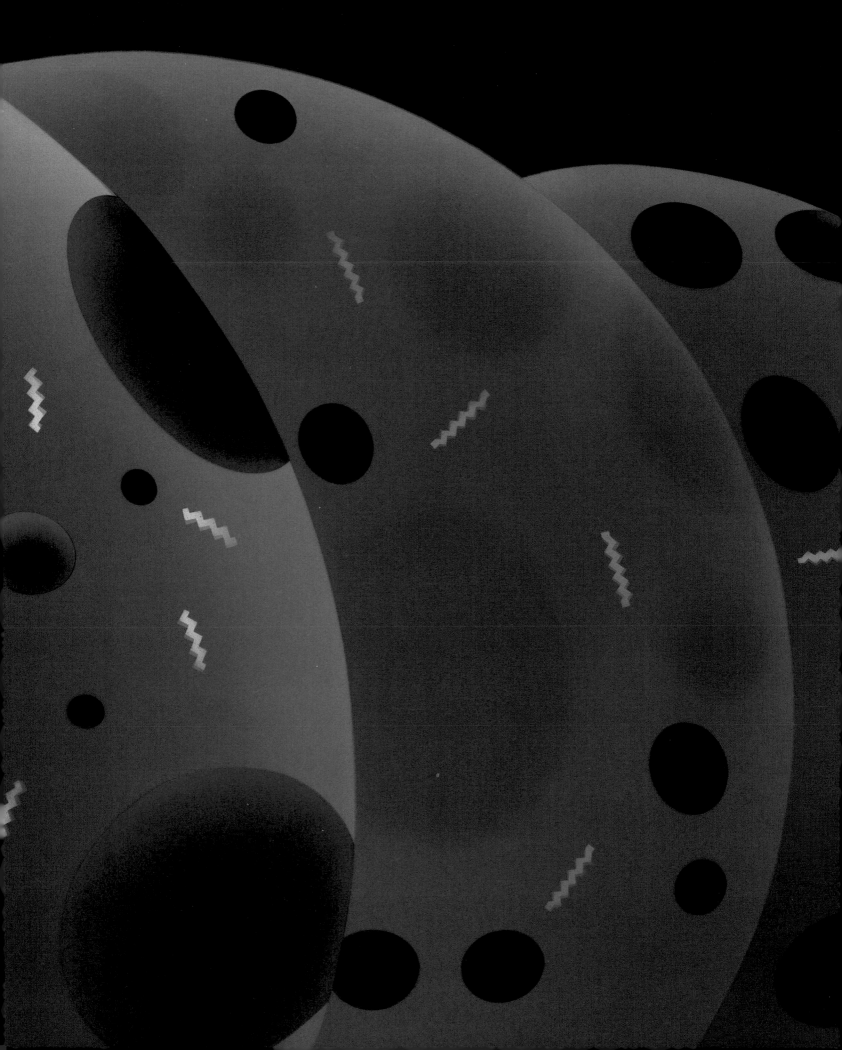

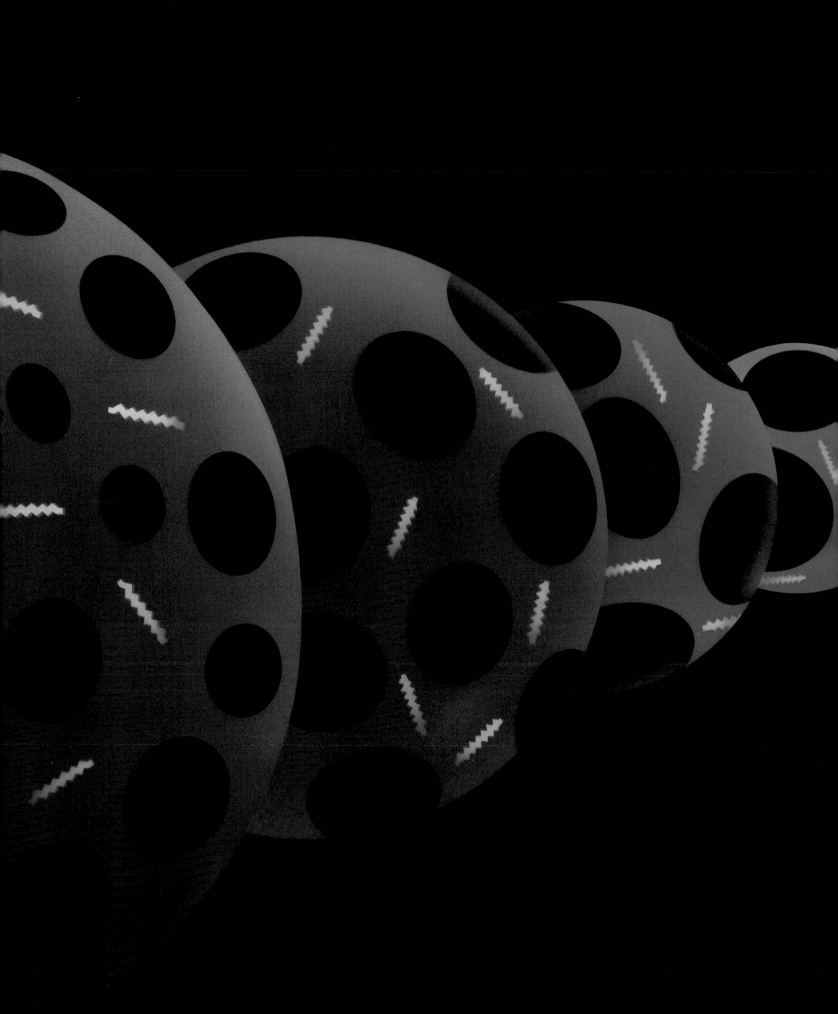

Into Nothingness?

In the final hours before the Big Crunch, the universe will bear a certain resemblance to itself in infancy. The major exception will be its multitudes of supermassive black holes. Drawn together by the now rapid implosion of space, the black holes will begin to cannibalize each other, growing even more monstrous.

As the contraction continues *(left),* the temperature of the intensely blue-shifted radiation—photons left over from the Big Bang and x-rays and gamma rays *(white)* from the combustion of ancient stars—will escalate to 10^{12} degrees Kelvin. At this point, protons and neutrons will disintegrate into their constituents, known as quarks. Then, as in a movie of the creation that is run in reverse, the distinctions between the fundamental forces of nature will begin to blur. Quarks will collide at phenomenal energies with other exotic subatomic fragments and produce more massive particles.

Feasting on this rich particle stew and on each other, black holes will begin to coalesce into one mammoth gravitational sink. Gravity will unite with the electromagnetic and strong and weak nuclear forces to create a "grand unified force," while the infinitesimal universe—now at 10^{32} degrees Kelvin—will be swallowed by the now ubiquitous black hole. As the laws of physics that govern the behavior of the known universe break down, scientists question whether the concept of a universe has any meaning at all.

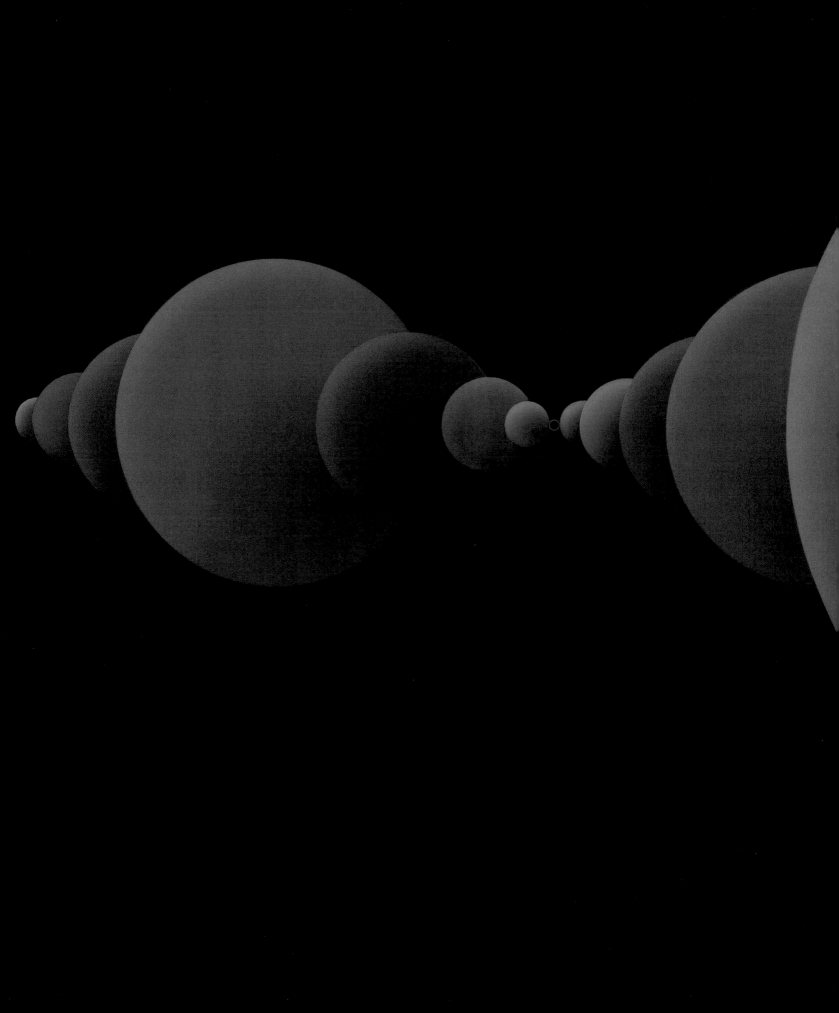

THE UNIVERSE REBOUNDS

Uncertainty over what takes place during the final instant in the life of the closed universe has led scientists to propose a poetic, if highly speculative, ending. Just before vanishing in a dimensionless point of infinite density, the cosmos may rebound *(left)*. Like its progenitor, the new universe will expand, spawning stars and galaxies and superclusters, and ultimately, the subatomic products of decay—until once again it reverses itself, collapsing to rebirth.

If the universe does oscillate—and, as yet, no driving mechanism has been identified—cosmologists suggest that it will increase in size and longevity with each cycle. The rationale for this theory centers on the photon. A high-energy photon emitted during the expansion phase of the closed universe will be stretched to longer and longer wavelengths by the outward rush of space, becoming red-shifted to the low-energy end of the spectrum.

When the contraction phase begins, that same photon will be compressed and its wavelength blue-shifted to an ever-higher energy state. At some point during cosmic collapse, the photon's wavelength will become shorter than it was at first issue; in effect, it will gain energy. The net energy of the shrinking cosmos will therefore also rise. If this energy is conserved through the rebound, the next universe will stay hotter longer and expand farther.

Appealing as it is, the theory is far from perfect. Scientists have yet to explain why or how a bounce might occur, or how a universe lumpy with black holes could give rise to one as relatively homogeneous as the current one. Despite such misgivings, theoretical physicists have calculated the number of cycles that could have preceded the present one: about 100, based on the estimated maximum size of the current universe. Before that, presumably, there was an initial Big Bang; earlier still, as far as anyone can guess, there was nothingness. And so it goes: Even as theorists wrestle with the uncertainties surrounding the fate of the cosmos, they ultimately confront the greatest mystery of all—the creation.

GLOSSARY

Absolute zero: the temperature, equal to about -459 degrees Fahrenheit (-273 degrees Celsius), at which molecular motion nearly ceases, and no thermal energy is obtainable.

Accretion disk: a disk formed from gases and other materials drawn in by a compact body, such as a black hole or neutron star, at the disk's center.

Active galactic nucleus (AGN): the center of a galaxy that exhibits energy not related to normal star processes. For example, quasars and Seyfert galaxies are classed as AGNs.

Antimatter: matter made up of antiparticles.

Antiparticle: a particle identical in mass to a matter particle, but opposite it in properties such as electrical charge. For example, a positron is the antiparticle to an electron.

Asteroid: a small, rocky, airless body that orbits a star.

Big Bang: according to a widely accepted theory, the primeval moment, 15 to 20 billion years ago, when the universe began expanding from a state of infinite density.

Big Bounce: term describing the theoretical behavior of a universe that, after contraction to almost infinite density, rebounds outward in rapid expansion.

Big Crunch: term used by some cosmologists to describe the final process of collapse theorized for a closed universe, when it contracts to a point of infinite density.

Black dwarf: the hypothetical remnant of a red, brown, or white dwarf star that has consumed its nuclear fuel.

Black hole: theoretically, an extremely compact body with such great gravitational force that nothing, not even radiation, can escape from within its event horizon. Some proposed varieties include stellar black holes, which form from the cores of very massive stars that have gone supernova; supermassive black holes, equivalent to several hundred million stars in mass and located in the centers of galaxies; and supergalactic black holes, formed from the merging of multiple galaxies.

Blueshift: a compression of the wavelengths of light which shifts their spectral lines toward the blue end of the spectrum. *See* Doppler effect, redshift.

Bow shock: the boundary region of interplanetary space where the solar wind is first deflected by a planet's magnetic field.

Carbonate rock cycle: the geochemical process in which carbon dioxide is removed from the atmosphere, deposited in rocks, and then, through weathering action, washed into the seas, where it either evaporates directly into the atmosphere, or is cycled through limestone for later release as volcanic gases.

Carbon dioxide: a compound of carbon and oxygen; also a so-called greenhouse gas in Earth's atmosphere that tends to trap heat at the planet's surface.

Cepheid variable: a star that changes regularly in luminosity over a set period of days or weeks.

Closed universe: a universe in which the average density of mass exceeds a critical value, so that gravity will eventually reverse the expansion of space.

Cluster: a gravitationally bound system of galaxies, ranging in number from a few to several thousand. A supercluster is a loose association of several galaxy clusters.

Comet: an asteroid-size body of dusty ice that travels in an elongated orbit around the Sun.

Core: the innermost region of a body. Earth's core is thought to consist of two irregular concentric spheres: an outer core of molten metal and a solid inner core believed to be mostly iron.

Cosmic microwave background: a steady microwave radiation detectable in all directions of the sky, commonly regarded as the cooled remnant of energy released in the instant of the Big Bang.

Cosmic ray: an atomic nucleus or charged particle moving at close to the speed of light; thought to originate in supernovae and other violent celestial phenomena.

Cosmological constant: a mathematical factor introduced by Einstein into the field equations of general relativity to accommodate his belief in a static universe. Today the cosmological constant is predicted by some theories, but is normally set at zero and thus disregarded.

Critical density: a crucial value for the density of matter in the universe, about 4.5 times 10^{-29} gram per cubic centimeter, or three hydrogen atoms per cubic meter. The relation between the actual mass density of the universe and the critical density determines whether the universe is closed, flat, or open.

Dark matter: a form of matter that has not yet been directly observed, but whose existence is deduced from its gravitational effects.

Density fluctuation: the unequal distribution of matter in the universe, sometimes termed "clumpiness."

Deuterium: a form of hydrogen having one neutron and one proton in its nucleus; also known as "heavy hydrogen."

Dipolar: having the characteristics of a magnetic dipole; that is, with two equal and opposite charges, positive and negative, in combination but separated by distance. The Earth's magnetic field is usually characterized as dipolar.

Doppler effect: a phenomenon in which waves appear to compress as their source approaches the observer (blueshift) or stretch out as the source recedes from the observer (redshift).

Dyson sphere: a hollow sphere constructed around a star by a hypothetical, technologically advanced society for the purpose of harnessing the star's energy. The concept of such spheres was suggested by physicist Freeman Dyson.

Electromagnetic radiation: waves of electrical and magnetic energy that travel through space at the speed of light.

Electromagnetism: the force that attracts oppositely charged particles and repels similarly charged particles. Electromagnetism does not affect neutral particles such as neutrinos.

Electron: a negatively charged particle that normally orbits an atom's nucleus but may exist in isolation.

Elliptical galaxy: a galaxy shaped like a flattened sphere, with no discernible interior structures and no disk.

Entropy: the measure of energy unavailable for doing work in a system or in the universe; increasing entropy implies a progression toward disintegration and randomness in the universe.

Ergosphere: the region around a black hole, outside the event horizon, where only objects that remain in motion can avoid entering the singularity.

Escape velocity (cosmic): the critical speed galaxies must attain in order to break free of their mutual gravitational attraction.

Event horizon: the spherical boundary around a black hole's singularity, within which gravitational forces prevent anything, including light, from escaping.

Exotic matter: theoretical particles invoked to explain certain observed effects of matter.

Fission: a nuclear reaction that releases energy when heavyweight atomic nuclei break down into lighter nuclei. *See* fusion.

Fusion: the combining of two atomic nuclei to form a heavier nucleus, releasing energy as a by-product.

Galactic plane: the central plane of the Milky Way galaxy; also, the central plane of any disk-shaped galaxy.

Galaxy: a system of stars, gas, and dust that contains millions to hundreds of billions of stars.

Gamma radiation (rays): the most energetic form of electromagnetic radiation, with the highest frequency and the shortest wavelength.

General relativity: a theoretical account of the effects of acceleration and gravity on the motion of bodies and the observed structure of space and time.

Giant: an aging, highly luminous star that has burned most of its hydrogen and expanded outward from its dense core.

Globular cluster: a spherical system of up to a few million stars that normally orbits the center of a spiral galaxy. Such clusters are thought to have formed early in a galaxy's evolution.

Grand unified theory (GUT): any of several competing but similar theories that unite electromagnetism, the weak force, and the strong force into one electronuclear force.

Gravity: the force responsible for the mutual attraction of separate masses.

Gravity wave: theoretically, radiation in the form of wavelike disturbances in an object's gravitational field that move energy away from their source. General relativity predicts that gravity waves may result from accelerating, oscillating, or violently disturbed masses.

Gravity well: a local distortion in the fabric of space-time near a massive body, analogous to a well or depression in a two-dimensional sheet.

Greenhouse effect: a phenomenon in which radiation is selectively transmitted and absorbed by gases in an atmosphere, admitting incoming short-wavelength solar radiation but blocking outgoing long-wavelength infrared, thus trapping heat near the surface of a planet.

Helium: the second lightest chemical element and the second most abundant, with a nucleus that includes two protons and at least two neutrons.

Hubble constant: an estimate of the rate at which the universe is expanding at the present time, based on a formula stating that widely separated objects recede from one another at a rate proportional to the distance between them. Values for the Hubble constant vary from 30 to 60 miles (50 to 100 kilometers) per second per megaparsec.

Hydrogen: the most common element in the universe. Stellar energy comes primarily from the fusion of hydrogen nuclei.

Ice age: a period during which polar ice sheets extend farther toward the equator, sometimes covering much of the planet's land areas with glaciers.

Inflationary cosmology: a theory describing the development of the universe that includes a sudden expansion of the universe occurring 10^{-35} second after the Big Bang.

Infrared: a band of radiation with a lower frequency and a longer wavelength than visible light.

Interglacial: relatively short periods between ice ages when glaciers retreat and temperatures rise.

Interstellar medium: the space between the stars. Though largely empty, it contains clouds of dust and gas at varying densities.

Inverse-square law: the mathematical relationship that describes the change in brightness of a star, or other point source of light, that occurs in inverse proportion to the square of the distance from the source; also any similar mathematical formula that describes how certain forces, such as gravity, change in strength with distance from a central point.

Iridium layer: a layer of clay dating from 66 million years ago that contains high levels of iridium, an element that is rare on Earth but relatively common in asteroids. Because the clay deposits have been found in many parts of the world, they may be evidence of an enormous asteroid collision that occurred at a time coinciding with a period of mass extinctions.

Irregular galaxy: a small galaxy of no classifiable structure or symmetry.

Kelvin: an absolute temperature scale that uses Celsius degrees but sets 0 at absolute zero, or about -273 degrees Celsius.

Kinetic energy: an object's energy of motion.

Lepton: a subatomic particle that is unaffected by the strong nuclear force; leptons include electrons, neutrinos, and others.

Leptoquark: according to theory, a massive subatomic particle that enables leptons and quarks to exchange identities.

Light-year: an astronomical unit of distance equal to the distance light travels in a vacuum in one year, or almost six trillion miles.

Luminosity: an object's total energy output, usually measured in ergs per second.

Magma: molten material formed deep within a planet or moon that may force its way through the crust to the surface, where it cools as lava.

Magnetic field: the area around a magnet, an electrical current, or a charged particle in which magnetic influence is felt by other currents, fields, and particles. Planetary magnetic fields, like those of a simple bar magnet, exhibit north and south poles linked by lines of varying magnetic strength and direction.

Magnetic field lines (flux): lines of force that indicate the strength and direction of flow in a magnetic field.

Magnetosphere: a large, energetic envelope of magnetic field lines shaped by interactions between a planet's magnetic field and the solar wind, the flow of charged particles from the Sun.

Magnetotail: a billowing, streamlike extension of a planet's magnetosphere formed on the planet's dark side by the action of the solar wind.

Mantle: the layer that extends from beneath Earth's crust approximately 1,800 miles to the core.

Nebula: a cloud of interstellar gas and dust; in some cases a supernova remnant or a shell ejected by a star. *See* Planetary nebula.

Neutrino: a chargeless particle with little or no mass that moves at or close to the speed of light.

Neutron: an uncharged particle with a mass similar to that of a proton; normally found in an atom's nucleus.

Neutron star: a very dense body composed of tightly packed neutrons; one possible product of a supernova explosion. Neutron stars are sometimes observed as pulsars.

Oort cloud: in astronomical theory, the largest and most distant of three contiguous cometary reservoirs. Named for its proposer, Dutch astronomer Jan Oort, it is envisioned as a huge, spherical cloud that surrounds the Solar System and extends some five trillion miles from the Sun.

Open universe: a universe that is on average less dense than the critical density and so continues to expand.

Ozone: the highly reactive, unstable three-atom form of the element oxygen, which normally occurs as a two-atom molecule. Near the surface, ozone is a toxic pollutant; in the stratosphere, it absorbs incoming solar ultraviolet rays that are harmful to life on the surface.

Parallax: a star's apparent motion on the celestial sphere over a six-month period. Measured in seconds of arc, it is used to determine a star's distance; the greater the parallax, the nearer the star.

Particle: the smallest component of any class of matter; for example, the elementary particles within an atom (such as electrons, protons, and neutrons); the smallest constituents of a gas (atoms and molecules); or the smallest forms of solid matter in space (interstellar and interplanetary dust particles).

Particle decay: the spontaneous transformation of a particle into one or more other particles, which may then decay as well.

Photon: a unit of electromagnetic energy associated with a specific wavelength. It behaves as a chargeless particle traveling at the speed of light.

Pion: an elementary particle that combines in triplets to produce particles having either positive, negative, or neutral charge.

Pi zero particle: a pion with a neutral charge.

Planetary nebula: a shell of gas, ejected from an aging red giant star.

Plasma: a gas of ionized particles, in contrast to ordinary gases, which are electrically neutral. Plasmas are sensitive to electrical and magnetic fields and are considered to be a fourth state of matter, along with ordinary gases, liquids, and solids.

Positron: an antiparticle to the electron, carrying a positive electric charge.

Positronium: an atomlike formation, composed of an electron and a positron, that emits radiation as it decays.

Protogalaxy: a roughly spherical hydrogen cloud from which a galaxy forms; about thirty times the size of a mature galaxy.

Proton: a positively charged particle, normally found in an atom's nucleus, with 1,836 times the mass of an electron.

Pulsar: a radiating source that emits extremely regular bursts of energy at intervals of several seconds or less. Pulsars are almost certainly neutron stars.

Quadrupole field: a magnetic field characterized by a combination of two dipoles of equal but opposite magnetic force.

Quantum: a fixed packet of some physical property, such as mass or energy.

Quantum mechanics: a mathematical description of the rules by which subatomic particles interact, decay, and form atomic or nuclear objects.

Quantum tunneling: a phenomenon invoked to explain the movement of subatomic particles through otherwise impenetrable force barriers, such as into or out of an atomic nucleus.

Quark: an elementary structural unit that combines to form particles including protons, neutrons, and pions.

Quasar: shortened from "quasi-stellar radio source"; an extremely powerful, bright source of energy, located in a very small region in the center of a distant galaxy, that outshines the whole galaxy around it.

Radiation: energy in the form of electromagnetic waves or subatomic particles.

Recessional velocity: the speed at which a body, particularly a galaxy, moves away from an observer; in general, more distant bodies appear to be moving more rapidly.

Red dwarf: a dim, long-lived, low-mass star.

Red giant: an aging, low-mass star that has expanded and cooled after consuming most of its core hydrogen.

Redshift: a stretching of the wavelengths of light, which shifts their spectral lines toward the red end of the spectrum. A Doppler redshift is caused by the motion of the light source; a cosmological redshift, by the expansion of space between the observer and the light source; and a gravitational redshift, by the time-distorting effects of the gravity of massive bodies.

Relativity: *see* General relativity.

Ring galaxy: a rare form of galaxy, thought to have been caused by the passage of a small galaxy through a larger disk galaxy, pulling the disk's stars into a ringlike formation.

Seyfert galaxy: an active disk galaxy with a very bright, starlike nucleus.

Singularity: the infinitely condensed mass at the center of a black hole. A singularity has no dimensions.

Solar cosmic rays: charged particles, usually protons or atomic nuclei of the light elements, emitted by the Sun.

Solar flare: an explosive release of charged particles and electromagnetic radiation from a small area on the surface of the Sun.

Solar wind: a continuous current of charged particles that streams outward from the Sun through the Solar System.

Space-time: a four-dimensional concept of the universe that incorporates three spatial dimensions plus time.

Spectrum: the array of electromagnetic radiation, arranged in order of wavelength, from long-wave radio to short-wave gamma rays. Also, a narrow band of wavelengths, such as the visible spectrum, in which light dispersed by a prism or other means shows its component colors; often banded with emission or absorption lines.

Spiral galaxy: a disk galaxy with bright stars and ionized gas clouds that form a pattern of two or more spiral arms curving out from the galaxy's center.

Spiral nebulae: distant galaxies once thought to be patchy star clusters within the Milky Way. Shifts in the spectral lines of these galaxies helped to establish the concept of an expanding universe.

Strong nuclear force: the force that binds quarks together into composite particles and holds protons and neutrons together to form atomic nuclei.

Subatomic particle: any particle smaller than an atom, from atomic components such as protons to the constituents of protons, quarks.

Subduction zone: the region where a spreading seafloor plate, for example, is forced under a continental one, or subducted. Subduction occurs mainly along the deep ocean trenches, where crustal material is driven back into the mantle.

Supergiant: an old, high-mass star greatly expanded from its original size; larger and brighter than a giant star.

Supernova: a stellar explosion that expels all or most of a star's mass and is extremely luminous.

Tectonic plate: a rigid segment of Earth's outer surface which moves in relation to other plates atop the mantle; movement occurs over periods of millions of years. There are about ten such plates.

Ultraviolet: a band of electromagnetic radiation with a higher frequency and shorter wavelength than visible blue light.

Virtual particles: extremely short-lived particles created out of nothingness, as permitted by the uncertainty principle. Although they exist too briefly to be directly observed, the effects of their existence may be detected.

Wavelength: the distance from crest to crest or trough to trough of an electromagnetic or other wave. Wavelengths are related to frequency; the longer the wavelength, the lower the frequency.

Weak nuclear force: a very short-range force responsible for particle decay.

White dwarf: an old, extremely dense star whose core has collapsed after exhausting its nuclear fuel of hydrogen and helium. A white dwarf with the mass of the Sun would be about the size of the Earth.

X-rays: a band of electromagnetic radiation intermediate in wavelength between ultraviolet radiation and gamma rays.

BIBLIOGRAPHY

Books

Abell, George O. *Exploration of the Universe* (4th ed.). Philadelphia: Saunders College, 1982.

Arp, Halton. *Quasars, Redshifts, and Controversies.* Berkeley, Calif.: Interstellar Media, 1987.

Audouze, Jean, and Guy Israël (eds.). *The Cambridge Atlas of Astronomy.* Cambridge, England: Cambridge University Press, 1985.

Bath, Geoffrey (ed.). *The State of the Universe: Wolfson College Lectures 1979.* Oxford, England: Clarendon Press, 1980.

Berger, A. (ed.). *The Big Bang and Georges Lemaître.* Dordrecht, Netherlands: D. Reidel, 1984.

Bernstein, Jeremy, and Gerald Feinberg (eds.). *Cosmological Constants: Papers in Modern Cosmology.* New York: Cambridge University Press, 1986.

Boslough, John. *Stephen Hawking's Universe.* New York: Quill/William Morrow, 1985.

Brown, Bruce, and Lane Morgan. *The Miracle Planet.* New York: Gallery Books, 1990.

Campbell, Joseph. *The Mythic Image.* Princeton, N.J.: Princeton University Press, 1974.

Chapman, Clark R., and David Morrison. *Cosmic Catastrophes.* New York: Plenum Press, 1989.

Close, Frank. *Apocalypse When?: Cosmic Catastrophe and the Fate of the Universe.* New York: William Morrow, 1988.

Close, Frank, Michael Marten, and Christine Sutton. *The Particle Explosion.* New York: Oxford University Press, 1987.

Clube, Victor, and Bill Napier. *The Cosmic Serpent: A Catastrophist View of Earth History.* New York: Universe Books, 1982.

Contopoulos, G., and D. Kotsakis. *Cosmology: The Structure and Evolution of the Universe.* Berlin: Springer-Verlag, 1986.

Darling, David. *Deep Time.* New York: Delacorte Press, 1989.

Davies, Paul:
The Accidental Universe. Cambridge, England: Cambridge University Press, 1982.
The Physics of Time Asymmetry. Berkeley: University of California Press, 1974.

Davies, Paul (ed.). *The New Physics.* Cambridge, England: Cambridge University Press, 1989.

Dyson, Freeman J. *Infinite in All Directions.* New York: Harper & Row, 1988.

Field, George, Gerrit L. Verschuur, and Cyril Ponnamperuma. *Cosmic Evolution: An Introduction to Astronomy.* Boston: Houghton Mifflin, 1978.

Forward, Robert L. *Future Magic.* New York: Avon Books, 1988.

Friedman, Herbert. *The Amazing Universe.* Washington, D.C.: National Geographic Society, 1975.

Fritzsch, Harald. *The Creation of Matter: The Universe from Beginning to End.* New York: Basic Books, 1984.

Gamow, George. *Thirty Years that Shook Physics: The Story of Quantum Theory.* New York: Dover Publications, 1966.

Godart, O., and M. Heller. *Cosmology of Lemaître.* Tucson, Ariz.: Pachart, 1985.

Gribbin, John:
In Search of the Big Bang: Quantum Physics and Cosmology. Toronto: Bantam Books, 1986.
White Holes: Cosmic Gushers in the Universe. New York: Dell, 1977.

Hawking, Stephen W. *A Brief History of Time: From the Big Bang to Black Holes.* Toronto: Bantam Books, 1988.

Henbest, Nigel. *The Exploding Universe.* New York: Macmillan, 1979.

Hodge, Paul W. *The Physics and Astronomy of Galaxies and Cosmology.* New York: McGraw-Hill, 1966.

Hoffman, Banesh, and Helen Dukas. *Albert Einstein: Creator and Rebel.* New York: Viking Press, 1972.

Hoyle, Fred. *Ten Faces of the Universe.* San Francisco: W. H. Freeman, 1977.

Ice Ages (Planet Earth series). Alexandria, Va.: Time-Life Books, 1983.

Islam, Jamal N. *The Ultimate Fate of the Universe.* Cambridge, England: Cambridge University Press, 1983.

Jacobs, J. A. *Reversals of the Earth's Magnetic Field.* Bristol, England: Adam Hilger, 1984.

Kaler, James B. *Stars and Their Spectra.* Cambridge, England: Cambridge University Press, 1989.

Kaufmann, William J., III:
Discovering the Universe. New York: W. H. Freeman, 1987.
Relativity and Cosmology. New York: Harper & Row, 1973.
Relativity and Cosmology (2d ed.). New York: Harper & Row, 1977.
Universe (2d ed.). New York: W. H. Freeman, 1987.

Kippenhahn, Rudolf. *Light from the Depths of Time.* Berlin: Springer-Verlag, 1984.

Kopal, Zdeněk. *Man and His Universe.* New York: William Morrow, 1972.

Krupp, E. C. *Echoes of the Ancient Skies: The Astronomy of Lost Civilizations.* New York: Harper & Row, 1983.

Merrill, Ronald T., and Michael W. McElhinny. *The Earth's Magnetic Field: Its History, Origin and Planetary Perspective.* London: Academic Press, 1983.

Misner, Charles W., Kip S. Thorne, and John Archibald Wheeler. *Gravitation.* San Francisco: W. H. Freeman, 1973.

Morris, Richard. *The End of the World.* Garden City, N.Y.: Anchor Press/Doubleday, 1980.

Motz, Lloyd. *The Universe: Its Beginning and End.* New York: Charles Scribner's Sons, 1975.

Narlikar, Jayant:
The Structure of the Universe. Oxford, England: Oxford University Press, 1977.
Violent Phenomena in the Universe. Oxford, England: Oxford University Press, 1982.

Nicolson, Iain. *Gravity, Black Holes and the Universe.* New York: John Wiley & Sons, 1981.

O'Flaherty, Wendy Doniger (ed.). *Textual Sources for the Study of Hinduism.* Totowa, N.J.: Barnes & Noble, 1985.

Parker, Barry. *Invisible Matter and the Fate of the Universe.* New York: Plenum Press, 1989.

Press, Frank, and Raymond Siever. *Earth* (2d ed.). San Francisco: W. H. Freeman, 1978.

Prokhovnik, S. J. *Light in Einstein's Universe: The Role of Energy in Cosmology and Relativity.* Dordrecht, Netherlands: D. Reidel, 1985.

Raup, David M. *The Nemesis Affair: A Story of the Death of Dinosaurs and the Ways of Science.* New York: W. W. Norton, 1986.

Rees, Martin, Remo Ruffini, and John Archibald Wheeler. *Black Holes, Gravitational Waves and Cosmology: An Introduction to Current Research.* New York: Gordon and Breach Science Publishers, 1974.

Sagan, Carl. *Cosmos.* New York: Random House, 1980.

Sagan, Carl, and Ann Druyan. *Comet.* New York: Random House, 1985.

Sayen, Jamie. *Einstein in America: The Scientist's Conscience in the Age of Hitler and Hiroshima.* New York: Crown, 1985.

Sciama, D. W. *Modern Cosmology.* Cambridge, England: Cambridge University Press, 1971.

Silk, Joseph. *The Big Bang: The Creation and Evolution of the Universe.* San Francisco: W. H. Freeman, 1980.

Smith, David G. (ed.). *The Cambridge Encyclopedia of Earth Sciences.* Cambridge, England: Cambridge University Press, 1981.

Smoluchowski, Roman, John N. Bahcall, and Mildred S. Matthews (eds.). *The Galaxy and the Solar System.* Tucson: University of Arizona Press, 1986.

Sullivan, Walter. *Black Holes: The Edge of Space, the End of Time.* Garden City, N.Y.: Anchor Press/Doubleday, 1979.

Trefil, James S. *Space Time Infinity.* New York: Pantheon Books, 1985.

Tucker, Wallace, and Karen Tucker. *The Dark Matter.* New York: William Morrow, 1988.

Unsöld, Albrecht, and Bodo Baschek. *The New Cosmos* (3d ed.). Berlin: Springer-Verlag, 1983.

Verschuur, Gerrit L. *Cosmic Catastrophes.* Reading, Mass.: Addison-Wesley Publishing, 1978.

Volcano (Planet Earth series). Alexandria, Va.: Time-Life Books, 1982.

Wald, Robert M. *Space, Time, and Gravity: The Theory of the Big Bang and Black Holes.* Chicago: University of Chicago Press, 1977.

Weinberg, Steven. *The First Three Minutes.* New York: Basic Books, 1977.

Whitrow, G. J. *The Structure and Evolution of the Universe: An Introduction to Cosmology.* New York: Harper & Brothers, 1959.

Will, Clifford M. *Was Einstein Right?* New York: Basic Books, 1986.

Zeilik, Michael, and John Gaustad. *Astronomy: The Cosmic Perspective.* New York: Harper & Row, 1983.

Periodicals

Abbott, Larry. "The Mystery of the Cosmological Constant." *Scientific American,* May 1988.

Abell, George O. "Cosmology—The Origin and Evolution of the Universe." *Mercury,* May-June 1978.

Adair, Robert K. "A Flaw in a Universal Mirror." *Scientific American,* February 1988.

Akasofu, Syun-Ichi. "The Dynamic Aurora." *Scientific American,* May 1989.

Bahcall, John N., Puragra Guhathakurta, and Donald P. Schneider. "What the Longest Exposures from the Hubble Space Telescope Will Reveal." *Science,* April 13, 1990.

Balick, Bruce. "The Shaping of Planetary Nebulae." *Sky & Telescope,* February 1987.

Barnes, Joshua E. "Making Ellipticals by Mergers." *Nature,* March 29, 1990.

Bartusiak, Marcia. "If You Like Black Holes, You'll Love Cosmic Strings." *Discover,* April 1988.

Begelman, M. C., and M. J. Rees. "Can Cosmic Clouds Cause Climatic Catastrophes?" *Nature,* May 27, 1976.

Begley, Sharon. "The Big-Bubble Theory." *Newsweek,* June 7, 1982.

Bekenstein, Jacob D. "Extraction of Energy and Charge from a Black Hole." *Technical Physical Review D,* February 15, 1973.

Bloxham, Jeremy, and David Gubbins. "The Evolution of the Earth's Magnetic Field." *Scientific American,* December 1989.

Burbidge, Geoffrey. "Quasars, Redshifts, and Controversies." *Sky & Telescope,* January 1988.

Clark, D. H., W. H. McCrea, and F. R. Stephenson. "Frequency of Nearby Supernovae and Climatic and Biological Catastrophes." *Nature,* January 27, 1977.

Coleman, Sidney. "The 1979 Nobel Prize in Physics." *Science,* December 14, 1979.

Corwin, Mike, and Dale Wachowiak. "Discovering the Expanding Universe." *Astronomy,* February 1985.

Cox, Allan, G. Brent Dalrymple, and Richard R. Doell. "Reversals of the Earth's Magnetic Field." *Scientific American,* February 1967.

Crutzen, Paul J., Ivar S. A. Isaksen, and George C. Reid. "Solar Proton Events: Stratospheric Sources of Nitric Oxide." *Science,* August 8, 1975.

Darling, David:
"Deep Time: The Fate of the Universe." *Astronomy,*

January 1986.

"Quest for Black Holes." *Astronomy,* July 1983.

Davies, Paul:

"Does the Universe Rotate?" *Sky & Telescope,* June 1988.

"Matter-Antimatter." *Sky & Telescope,* March 1990.

Davis, Marc, Piet Hut, and Richard A. Muller. "Extinction of Species by Periodic Comet Showers." *Nature,* April 19, 1984.

"Deep Space." *Sky & Telescope,* November 1989.

Dietz, Robert S., and John C. Holden. "The Breakup of Pangaea." *Scientific American,* October 1970.

Dicus, Duane A., et al. "The Future of the Universe." *Scientific American,* March 1983.

Dyson, Freeman J.:

"Reflections: Disturbing the Universe—1." *The New Yorker,* August 6, 1979.

"Reflections: Disturbing the Universe—2." *The New Yorker,* August 13, 1979.

"Reflections: Disturbing the Universe—3." *The New Yorker,* August 20, 1979.

"Time Without End: Physics and Biology in an Open Universe." *Review of Modern Physics,* July 1979.

Fienberg, Richard Tresch. "The New, Improved Space Telescope." *Sky & Telescope,* February 1989.

Gale, George. "The Anthropic Principle." *Scientific American,* December 1981.

Gamow, George. "The Evolutionary Universe." *Scientific American,* September 1956.

Gore, Rick. "The Once and Future Universe." *National Geographic,* June 1983.

Gott, J. Richard, III, et al. "Will the Universe Expand Forever?" *Scientific American,* March 1976.

Griggs, Tony. "Researchers Team Up to Ponder Weight—and Future—of Universe." *Bell Labs News,* October 7, 1985.

Guth, Alan H., and Paul J. Steinhardt. "The Inflationary Universe." *Scientific American,* May 1984.

Gwynne, Peter. "Physics: Unifying the Fields." *Newsweek,* October 29, 1979.

Hawking, Stephen W. "The Quantum Mechanics of Black Holes." *Scientific American,* January 1977.

Hoffman, Kenneth A. "Ancient Magnetic Reversals: Clues to the Geodynamo." *Scientific American,* May 1988.

Hoffman, Kenneth A., and M. Fuller. "Transitional Field Configurations and Geomagnetic Reversal." *Nature,* June 29, 1978.

Jeanloz, Raymond. "The Earth's Core." *Scientific American,* September 1983.

Jones, Brian. "The Legacy of Edwin Hubble." *Astronomy,* December 1989.

"In Search of the Oldest Galaxies." *Science News,* August 6, 1988.

Islam, Jamal N. "The Ultimate Fate of the Universe." *Sky & Telescope,* January 1979.

Kanipe, Jeff:

"Galaxies at the Confusion Limit." *Astronomy,* December 1988.

"Quest for the Most Distant Objects in the Universe." *Astronomy,* June 1988.

"Too Smooth: COBE's Perfect Universe." *Astronomy,* June 1990.

Kasting, James F., Owen B. Toon, and James B. Pollack.

"How Climate Evolved on the Terrestrial Planets." *Scientific American,* February 1988.

Keel, William C. "Crashing Galaxies, Cosmic Fireworks." *Sky & Telescope,* January 1989.

Killian, Anita M. "Playing Dice with the Solar System." *Sky & Telescope,* August 1989.

Kippenhahn, Rudolf. "Light from the Depths of Time." *Sky & Telescope,* February 1987.

King, Ivan R. "Globular Clusters." *Scientific American,* June 1985.

Lanzerotti, Louis J. "Earth's Magnetic Environment." *Sky & Telescope,* October 1988.

LoSecco, J. M., Frederick Reines, and Daniel Sinclair. "The Search for Proton Decay." *Scientific American,* June 1985.

McCrea, W. H. "Ice Ages and the Galaxy." *Nature,* June 19, 1975.

Mallove, Eugene F. "The Self-Reproducing Universe." *Sky & Telescope,* September 1988.

Mann, Charles C., and Robert P. Crease. "Waiting for Decay." *Science 86,* March 1986.

March, Robert. "Protons Are Not Forever." *Omni,* November 1980.

Marschall, Laurence A., Liang-Tai George Chiu, and William F. van Altena. "Star Cluster Membership: Separating Sheep from Goats." *Sky & Telescope,* August 1981.

Montes, Augusto F. Molina. "The Building of Tenochtitlan." *National Geographic,* December 1980.

"Most Distant Object in Universe Discovered." *Astronomy,* June 1988.

"Mysterious Objects May Be Youngest Galaxies." *Astronomy,* April 1988.

Nather, R. Edward. "Learning from White Dwarfs." *Star Date,* May-June 1990.

"Old Galaxy, Young Universe." *Astronomy,* September 1988.

Osmer, Patrick S. "Quasars as Probes of the Distant and Early Universe." *Scientific American,* February 1982.

Overbye, Dennis:

"Exploring the Edge of the Universe." *Discover,* December 1982.

"The Universe According to Guth." *Discover,* June 1983.

"Why He Ranks as a World Class Scientist." *Time,* May 14, 1990.

Page, Don N., and M. Randall McKee. "The Future of the Universe." *Mercury,* January-February 1983.

Parker, Barry. "The Mysterious Dark Matter." *Star Date,* November-December 1989.

Penrose, Roger. "Black Holes." *Scientific American,* May 1972.

Peratt, Anthony L. "Not with a Bang." *The Sciences,* January-February 1990.

Peterson, I. "Out-of-This-World View of the Milky Way." *Science News,* April 28, 1990.

Price, Richard H., and Kip S. Thorne. "The Membrane Paradigm for Black Holes." *Scientific American,* April 1988.

"Quasars Triggered by Galactic Collisions." *Astronomy,* May 1988.

Rampino, Michael R., and Richard B. Stothers. "Geological Rhythms and Cometary Impacts." *Science,* December 21, 1984.

Reid, G. C., et al. "Influence of Ancient Solar-Proton Events on the Evolution of Life." *Nature,* January 22, 1976.

Robinson, Leif J. "Galactic Cannibalism." *Sky & Telescope,* February 1981.

Rothman, Tony. "Time." *Discover,* February 1987.

Schechter, Bruce. "Search and Discover: Measuring the Hubble Constant with Supernovas." *Physics Today,* January 1986.

Schwartz, Richard D., and Philip B. James. "Periodic Mass Extinctions and the Sun's Oscillation about the Galactic Plane." *Nature,* April 19, 1984.

Schwarzschild, Bertram M. "Proton Decay Not Seen at Predicted Rate." *Physics Today,* September 1983.

Shu, Frank H. "The Expanding Universe and the Large-Scale Geometry of Spacetime." *Mercury,* November-December 1983.

Silk, Joseph. "Probing the Primeval Fireball." *Sky & Telescope,* June 1990.

Siscoe, G. L., and N. U. Crooker. "Auroral Zones in a Quadrupole Magnetosphere." *Journal of Geomagnetism and Geoelectricity,* 1976, Vol. 28, no. 1, pp. 1-9.

Siscoe, G. L., C. K. Chen, and M. Harel. "On the Altitude of the Magnetopause During Geomagnetic Reversals." *Journal of Atmospheric and Terrestrial Physics,* 1976, Vol. 38, pp. 1327-1331.

Stephenson, F. Richard, and David H. Clark. "Historical Supernovas." *Scientific American,* June 1976.

Sulentic, Jack W. "Are Quasars Far Away?" *Astronomy,* October 1984.

Thaddeus, Patrick, and Gary A. Chanan. "Cometary Impacts, Molecular Clouds, and the Motion of the Sun Perpendicular to the Galactic Plane." *Nature,* March 7, 1985.

Thorne, Kip S. "The Search for Black Holes." *Scientific American,* December 1974.

Trefil, James S. "Stop to Consider the Stones that Fall from the Sky." *Smithsonian,* September 1989.

Tucker, Wallace, and Karen Tucker. "Against All Odds: Matter and Evolution in the Universe." *Astronomy,* September 1984.

Verschuur, Gerrit L. "A New 'Yardstick' for the Universe." *Astronomy,* November 1988.

Waddington, C. J. "Paleomagnetic Field Reversals and Cosmic Radiation." *Science,* November 1967.

Waldrop, M. Mitchell:
"Collision and Cannibalism Shape the Galaxies." *Science,* February 3, 1989.
"Do-It-Yourself Universes." *Science,* January 1987.
"An Infinity of New Universes." *The Washington Post,* January 15, 1989.

Weinberg, Steven. "The Decay of the Proton." *Scientific American,* June 1981.

Wesson, Paul S. "Olbers' Paradox Solved at Last." *Sky & Telescope,* June 1989.

"White Dwarfs and the Age of the Universe." *Sky & Telescope,* October 1987.

Whitmire, Daniel, and Ray Reynolds. "The Fiery Fate of the Solar System." *Astronomy,* April 1990.

Other Sources

"New Ideas about Galaxy Formation." Press release. Washington, D.C.: American Astronomical Society, June 13, 1990.

"DIRBE Preliminary Sky Map at 240 Microns." Press release. Washington, D.C.: National Aeronautics and Space Administration, no date.

"Early COBE Results in Accord with Big Bang Theory." Press release. Washington, D.C.: National Aeronautics and Space Administration, January 13, 1990.

INDEX

ACKNOWLEDGMENTS

The editors wish to thank Halton Arp, Max Planck Institut für Physik und Astrophysik, Garching, Germany; Tanya Atwater, University of California, Santa Barbara; Nancy W. Boggess, NASA Goddard Space Flight Center, Greenbelt, Md.; John B. Carlson, Center for Archaeoastronomy, College Park, Md.; David H. Clark, Polaris House, Swindon, England; Thomas M. Dame, Harvard-Smithsonian Center for Astrophysics, Cambridge, Mass.; Stanislav Djorgovski, California Institute of Technology, Pasadena; Fraser P. Fanale, University of Hawaii, Manoa, Honolulu; Robert Futually, Observatoire Midi-Pyrénées, France; Maurice Goldhaber, Brookhaven National Laboratory, Upton, N.Y.; Paul Gorenstein, Harvard-Smithsonian Center for Astrophysics, Cambridge, Mass.; Richard Grieve, Geological Survey of Canada, Ottawa, Ontario; Puragra Guhathakurta, Princeton University, N.J.; Jonathan I. Lunine, University of Arizona, Tucson; Joseph M. Mazzarella, California Institute of Technology, Pasadena; Jacques-Clair Noëns, Observatoire du Pic-du-Midi, France; Peter Olson, Johns Hopkins University, Baltimore, Md.; Ann Palfreyman, California Institute of Technology, Pasadena; Michael R. Rampino, NASA Goddard Institute for Space Studies, New York; David M. Raup, University of Chicago, Ill.; François Schweizer, Carnegie Institute of Washington, Washington, D.C.; Neil R. Sheeley, Naval Research Laboratory, Washington, D.C.; George L. Siscoe, University of California, Los Angeles; Richard B. Stothers, NASA Goddard Institute for Space Studies, New York; Barbara Sweeney, AT&T Bell Laboratories, Short Hills, N.J.; Gerrit L. Verschuur, Bowie, Md.; Marie-Josée Vin, Observatoire de Haute Provence, France; Richard M. West, European Southern Observatory, Garching, Germany; Stanford E. Woosley, University of California, Santa Cruz.

PICTURE CREDITS

The sources for the illustrations in this book are listed below. Credits from left to right are separated by semicolons, from top to bottom by dashes.

Cover: Art by Alfred T. Kamajian. 6: Granger Collection. 7: Museo Nacional de Antropologia, Mexico City. 8: Initial cap, detail from page 7. 12, 13: Katia Krafft/Explorer, Paris. 14, 15: Art by Fred Holz, copied by Larry Sherer. 16: Naval Research Laboratory. 17: Art by Alfred T. Kamajian. 18, 19: Art by Alfred T. Kamajian; Bloxham, Gubbins & Jackson/Scientific American (2)—art by Karen Tuveson. 20, 21: Art by Alfred T. Kamajian and Karen Tuveson. 24, 25: Steve McCutcheon. 26, 27: Richard A. F. Grieve/Geological Survey of Canada, Ottawa (3)—David J. Roddy, Eugene M. Shoemaker, and Carolyn S. Shoemaker/U.S. Geological Survey, Flagstaff, Ariz. 30, 31: Art by Fred Holz, inset art by Alfred T. Kamajian, copied by Larry Sherer. 32, 33: Art by Fred Holz, copied by Larry Sherer. 35: Art by Fred Holz. 38: Dr. Bruce Balick, University of Washington. 41-53: Art by Rob Wood of Stansbury, Ronsaville, Wood, Inc.; inset art by Time-Life Books. 54, 55: NASA. 56: Initial cap, detail from pages 54, 55. 59: Courtesy AT&T Archives. 60, 61: Anthony Tyson and Puragra Guhathakurta, Institute for Advanced Study, Princeton, N.J. 63: Photo Researchers. 64, 65: Dr. Halton Arp, Max Planck Institut für Physik und Astrophysik, Garching, Germany. 66, 67: Art by Stephen Wagner. 68: Photo Researchers. 69: Courtesy Joseph Mazzarella, California Institute of Technology. 70: Computer-processed image based on a photographic plate taken at Cerro Tololo Inter-American Observatory by I. King (University of California at Berkeley) and S. Djorgovski (California Institute of Technology). 72, 73: Art by Stephen Wagner; from *Constructing the Universe* by David Layzer, courtesy MMI Corporation—Harvard-Smithsonian Center for Astrophysics—Dr. Bruce Balick, University of Washington. 75-77: Art by Stephen Wagner. 78, 79: Art by Fred Holz; National Optical Astronomy Observatories; Dr. Halton Arp, Max Planck Institut für Physik und Astrophysik, Garching, Germany; Dr. François Schweizer. 81: Art by Matt McMullen. 82-91: Art by Matt McMullen; inset art (bar graph) by Time-Life Books. 92, 93: NASA, Goddard Space Flight Center. 94: Initial cap, detail from pages 92, 93. 97, 98: Art by Fred Holz (2). 101: Palomar Observatory, California Institute of Technology. 103: Paul Gorenstein, Harvard-Smithsonian Center for Astrophysics. 104, 105: Background art by Al Kettler, courtesy National Portrait Gallery, London; Hebrew University of Jerusalem, courtesy AIP Niels Bohr Library; Leningrad Physico-Technical Institute, courtesy AIP Niels Bohr Library; UPI/Bettmann Newsphotos; Hale Observatories, courtesy AIP Niels Bohr Library; UPI/Bettmann Newsphotos. 106, 107: Background art by Al Kettler, Princeton University (2); courtesy AT&T Archives; Harvard University News Office (2); Mia Dyson; Donna Coveney, courtesy MIT. 110-113: Art by Alfred T. Kamajian (4). 116: David Parker/Photo Researchers—Irvine-Michigan-Brookhaven Proton Decay Experiment/Photo Researchers. 119-133: Art by Yvonne Gensurowsky of Stansbury, Ronsaville, Wood, Inc.

Time-Life Books Inc.
is a wholly owned subsidiary of
THE TIME INC. BOOK COMPANY

TIME-LIFE BOOKS INC.

Managing Editor: Thomas H. Flaherty
Director of Editorial Resources:
Elise D. Ritter-Clough
Director of Photography and Research:
John Conrad Weiser
Editorial Board: Dale Brown, Roberta Conlan,
Laura Foreman, Lee Hassig, Jim Hicks, Blaine
Marshall, Rita Mullin, Henry Woodhead

PUBLISHER: Joseph J. Ward

Associate Publisher: Trevor Lunn
Editorial Director: Donia Ann Steele
Marketing Director: Regina Hall
Director of Design: Louis Klein
Production Manager: Prue Harris
Supervisor of Quality Control: James King

Editorial Operations
Production: Celia Beattie
Library: Louise D. Forstall

Computer Composition: Deborah G. Tait
(Manager), Monika D. Thayer, Janet Barnes
Syring, Lillian Daniels

Correspondents: Elisabeth Kraemer-Singh (Bonn),
Christine Hinze (London), Christina Lieberman
(New York), Maria Vincenza Aloisi (Paris), Ann
Natanson (Rome). Valuable assistance was also
provided by Elizabeth Brown (New York), Andrea
Dabrowski (Mexico City), and Robert Kroon
(Genolier, Switzerland).

VOYAGE THROUGH THE UNIVERSE

SERIES EDITOR: Roberta Conlan
Series Administrator: Susan Stuck

Editorial Staff for *Frontiers of Time*
Art Director: Cynthia Richardson
Picture Editor: Tina McDowell
Text Editor: Robert M. S. Somerville
Associate Editor/Research: Mary H. McCarthy
Assistant Editors/Research: Mark Galan,
Patricia A. Mitchell
Writer: Darcie Conner Johnston
Assistant Art Director: Brook Mowrey
Copy Coordinator: Juli Duncan
Picture Coordinators: Jennifer Iker,
Barry Anthony
Editorial Assistant: Katie Mahaffey

Special Contributors: Sarah Brash, Andrew
Chaikin, George Constable, Ken Crosswell, John
Langone, David Lindley, Michael Lemonick, Peter
Pocock, Chuck Smith, M. Mitchell Waldrop (text);
Vilasini Balakrishnan, Mark Cheater, Jocelyn G.
Lindsay, Ted Loos (research); Barbara L. Klein
(index).

CONSULTANTS

GEOFFREY R. CHESTER has been a member of
the staff of the Smithsonian Institution's Albert
Einstein Planetarium in Washington, D.C. since
1978 and has lectured widely on all aspects of
astronomy. He is also a noted astrophotographer
whose work has appeared in several astronomy
magazines.

NANCY U. CROOKER, a professor of atmospheric
physics on leave from the University of California at
Los Angeles, is presently with the Geophysics Laboratory of Hanscom Air Force Base in Bedford,
Massachusetts, where she studies the interaction of
the solar wind with the terrestrial environment.

ALAN M. DRESSLER is an astronomer at the Carnegie Institution of Washington's Mount Wilson
and Las Campanas Observatories in Pasadena, California. He explores the formation and evolution of
galaxies.

KENNETH A. HOFFMAN is a professor of physics at
the California Polytechnic State University in San
Luis Obispo where he studies the process of geomagnetic reversal.

JAMES B. KALER, an expert on spectroscopy, teach-
es stellar astronomy at the University of Illinois,
Urbana.

RICHARD G. KRON is the director of the University
of Chicago's Yerkes Observatory. His interests include the evolution of galaxies and the detection of
very distant galaxies.

STEN ODENWALD is an infrared astronomer with
the Space Sciences Division of the Naval Research
Laboratory in Washington, D.C.

DON NELSON PAGE is a professor of physics at the
University of Alberta in Edmonton who researches
black holes and the early universe.

RICHARD PRICE, a professor of physics at the University of Utah, Salt Lake City, studies relativistic
astrophysics, the application of Einstein's theory of
gravitation to astrophysical phenomena particularly as it applies to black holes.

RAY T. REYNOLDS has been a planetary science
researcher for nearly thirty years at the Theoretical
Studies Branch of NASA Ames Research Center,
Moffett Field, California. His work focuses on the
formation, structure, and evolution of planets.

DOUGLAS O. RICHSTONE is a professor of astronomy at the University of Michigan, Ann Arbor. He
specializes in stellar dynamics, extragalactic astronomy, and cosmology.

DORIS ROSENBAUM TEPLITZ is a physicist at
Southern Methodist University in Dallas, Texas, re-
searching elementary particle theory and astrophysics. Her work on the future of the universe is
widely known.

VIGDOR L. TEPLITZ, chairman of the Physics De-
partment at Southern Methodist University, specializes in elementary particle theory and astrophysics. He spent several years at the U.S. Arms
Control and Disarmament Agency as a part of bi-
lateral negotiations with the Soviet Union.

DANIEL P. WHITMIRE, an astrophysicist at the
University of Southwestern Louisiana, investigates
the origin and evolution of planetesimal disks.

**Library of Congress Cataloging in
Publication Data**
Frontiers of time/by the editors of Time-Life
Books.
p.   cm.   (Voyage through the universe).
Bibliography: p.
Includes index.
ISBN 0-8094-6929-4.
ISBN 0-8094-6930-8 (lib. bdg.).
1. Cosmology.
I. Time-Life Books. II. Series.
QB981.F77   1991
523.1—dc20   90-11029 CIP

For information on and a full description of
any of the Time-Life Books series, please call
1-800-621-7026 or write:
Reader Information
Time-Life Customer Service
P.O. Box C-32068
Richmond, Virginia 23261-2068

Time-Life Books Inc. offers a wide range of fine
recordings, including a *Rock 'n' Roll Era* series.
For subscription information, call 1-800-621-7026
or write Time-Life Music, P.O. Box C-32068, Richmond, Virginia 23261-2068.

Earth: diameter 7,926 miles

Neptune: diameter 30,700 miles

Uranus: diameter 31,600 miles

Red supergiant: diameter 400 million miles

Solar System: diameter 7.5 billion miles

Globular cluster: diameter 2×10^{16} miles

Milky Way: diameter 100,000 light-years

Local Group of galaxies: 5 million light-years across

Largest double radio source: length 17 million light-years